CONVERSATIONS ON THE GEOMETRY OF REALITY

CONVERSATIONS ON THE GEOMETRY OF REALITY

A Critical Inquiry into the Foundations of Physics

Amir J. Guri, Ph.D., J.D.

NATURAL PHILOSOPHY PRESS · BROOMFIELD, COLORADO

Published by Natural Philosophy Press, Broomfield, Colorado

For ordering information and quantity discounts, contact Natural Philosophy Press.

Library of Congress Control Number: 2026937694

ISBNs: Hardcover 979-8-9941756-3-7 · Paperback 979-8-9941756-4-4 · eBook 979-8-9941756-5-1

First Edition · Printed in the United States of America

Why Dialogue, and How to Read It

This book follows *The Geometry of Reality*, a rigorous scientific monograph that introduces a new physical framework called Unified Field Dynamics (UFD). In *The Geometry of Reality* and its accompanying appendices, the aim was to present UFD in a scientifically and mathematically serious form and to show how it accounts for the empirical successes of modern physics while reducing ontological fragmentation and the number of free parameters.

In this book, the aim is different. Here, the framework is presented in dialogical form in order to address the reasonable concerns physicists may have about moving away from the highly successful models of modern physics, especially the Standard Model and General Relativity. The purpose is not to ask for agreement in advance. It is to put the framework through its paces in a form that makes its strengths, weaknesses, promises, and open burdens easier to see.

That choice of form belongs to an older tradition in natural philosophy. Galileo's *Dialogue Concerning the Two Chief World Systems* remains one of the best-known examples of how competing pictures of reality can be tested through sustained exchange rather than presented as settled conclusions from the start. By using dialogue to examine the emerging heliocentric view against the traditional geocentric one, Galileo allowed different temperaments, assumptions, and standards of explanation to confront one another so that the force of the argument could emerge through comparison and pressure rather than mere assertion. Although aim here differs in tone and scope from Galileo's work, the inspiration remains the same. Dialogue makes it possible to stage pressure, hesitation, clarification, and resistance in a way that ordinary exposition often cannot.

The book consists of a conversation between two physicists. Dr. S speaks from the standpoint of a serious mainstream physicist: cautious, technically minded, and unwilling to grant a new framework more than it has earned. Dr. U speaks from within UFD: committed to its ontology, responsible for stating it clearly, and equally responsible for not overstating what the framework has and has not yet closed. The point of this exchange is not theatrical contrast. It is disciplined pressure. Whenever possible, the dialogue begins with the most physically intuitive picture. Only when Dr. S presses for greater precision does Dr. U move to the deeper geometric structure or the more technical formulation. That layering is deliberate: intuition first, structure second, formalism third.

A note on claim maturity is especially important. Not every chapter in this book stands at the same level. Some sectors of UFD are presented as comparatively sharp benchmark domains with quantitative results. Others are presented as promising but incomplete. That unevenness is not a defect to be hidden; it is part of scientific honesty. A broad framework becomes more credible when it clearly marks where its strongest results lie and where its burdens remain. The distinction among derived claims, constrained claims, and programmatic claims — explained more fully in the Prologue — governs the entire conversation. Roughly speaking, a derived claim has been carried to quantitative closure, a constrained claim has an identified mechanism but not yet full closure, and a programmatic claim is best understood as a research direction rather than a completed result.

Readers may approach the book in more than one way. A general reader can move chapter by chapter, treating each conversation as a self-contained inquiry. A technically trained reader may prefer to watch for the recurring burdens: where UFD claims a fully derived result, where it offers a constrained mechanism that is not yet fully closed, and where it is still best understood as a programmatic research direction. A skeptical reader may simply follow Dr. S and ask, at each step, whether the vocabulary has been reduced to structural work.

Although this book grows out of *The Geometry of Reality* and its accompanying appendices, it is written to stand on its own. Each chapter begins with concise framing material so that readers can follow the central dispute without having to master the larger technical corpus first. All source materials are available at www.unifiedfielddynamics.com.

Whether you are a physicist who wants to see a new framework tested under pressure, or a general reader who wants a more physically intuitive picture of reality, I hope this exchange helps you judge the theory on its merits.

Methodological Note and Acknowledgments

Unified Field Dynamics arose from a philosophical project I had been developing for roughly ten years before the publication of *The Geometry of Reality*. Once I had formed a coherent philosophical blueprint of UFD, I used contemporary large language models — namely Gemini (Google), Claude (Anthropic), and ChatGPT (OpenAI) — to help develop its scientific articulation and, later, its mathematical formalization.

This sequence — philosophy, then science, then mathematics — is one of the features that gives the framework its unusual coherence and distinctiveness. Much contemporary physics proceeds in the opposite direction, beginning from mathematical formalism or empirical pattern and working outward toward interpretation. UFD began instead with the question of what kind of physical reality the world would have to be in order to hold together at all, and only then asked how that reality could be described scientifically and formalized mathematically. The result is a framework that aims to remain unified at the philosophical level even when the technical details become demanding.

This approach differs from the way large language models are often used in speculative theory generation. A common pattern is to prompt a model with a broad topic and ask it to produce an ambitious theory. The resulting output can sound impressive while lacking conceptual discipline, sustained authorship, or real responsibility for the structure being proposed. In such cases, the "theory" may exist mainly as a sequence of generated outputs rather than as a framework that has been critically developed and internally integrated by a human thinker.

UFD was not developed in that way. Its philosophical blueprint existed in draft form before any language model was consulted. When I later engaged with these models, I used them to test, extend, clarify, and formalize a framework I had already thought through at length, not to originate one. I could challenge outputs where they conflicted with the blueprint, revise them where they drifted, and reject material that did not fit the underlying architecture. The models functioned as research assistants and interlocutors rather than as oracles. Using multiple models at different stages also provided a useful comparative check: convergence sometimes identified lines of reasoning that appeared stable across systems, while divergence often revealed places where the framework required clarification or where alternative formal approaches were in tension.

That distinction matters because it bears directly on authorship. A theory generated only as model output has no author in the strongest intellectual sense: no one is fully responsible for its coherence, no one has absorbed its implications as a whole, and no one can defend it under sustained criticism. UFD has been developed, revised, defended, and rewritten by a single human author over more than a decade. The models contributed to later stages of that process. They did not begin it, and they do not conclude it.

In preparing this book, I used Claude (Anthropic) and ChatGPT (OpenAI) as auxiliary tools for drafting, comparative framing, editorial refinement, and structural experimentation. Their principal role was to assist in the clarification, presentation, and development of ideas whose underlying framework had already been established. All theoretical claims, interpretive judgments, and final editorial decisions remained my own.

CONTENTS

INTRODUCTION

A DIALOGUE IN SEARCH OF REALITY

The achievements of modern physics are not in doubt. It has measured, predicted, and organized an astonishing range of phenomena, from the behavior of atoms to the expansion history of the cosmos. Yet its success has not yielded a fully unified picture of what reality is. The deeper one looks, the more the prevailing landscape appears as a set of extraordinarily effective frameworks — gravity, quantum theory, nuclear structure, cosmology — each powerful in its own domain, yet only partially integrated with the others.

This book begins from that tension.

What follows is neither a textbook nor a staged debate. It is a structured conversation between two scientists. Dr. S speaks from within the mainstream inheritance: the Standard Model, general relativity, and modern cosmology. Dr. U speaks from within Unified Field Dynamics, or UFD. Each enters with real commitments. Neither begins from neutrality. But both accept a higher obligation than loyalty to a framework: they will follow the argument wherever it leads.

That matters because UFD is not offered as a small correction to existing theory. It is offered as a replacement-level framework built on different ontological assumptions: nested physical fields, stable topological excitations, coherence thresholds, and geometric boundary dynamics. Its claim is not merely that familiar equations can be redescribed in more intuitive language. Its claim is that matter, interaction, quantum behavior, cosmological structure, and related coordination phenomena may be manifestations of one deeper physical architecture.

A deeper issue runs beneath the formal standards. Modern physics has rightly taught restraint: calculate carefully, predict accurately, and do not demand that reality conform to familiar pictures. That restraint has often been a virtue. But it can harden into something weaker — the assumption that because a theory works, its opacity no longer matters; or that because nature is sometimes counterintuitive, it may therefore be permanently unintelligible in principle. This book proceeds from a different possibility.

Reality need not flatter intuition to reward understanding. The history of science suggests not that the world is absurd, but that its deeper order is often hidden beneath successful but provisional abstractions. The desire for a more intelligible picture is not childish. It is one of the recurring engines of scientific progress.

One methodological principle governs the mathematics throughout. UFD does not claim that the mathematical achievements of modern physics must be discarded. On the contrary, one of its central claims is that much of the mathematics may remain while the ontology changes. Continuity equations remain continuity equations. Wave equations remain wave equations. Boundary-value problems remain boundary-value problems. What changes is the physical interpretation of the quantities involved: fields become real media, particles become stable excitations, and many force-like phenomena become consequences of coherence, topology, pressure geometry, and boundary dynamics. The guiding rule is simple: ontology first, mechanism second, mathematics third.

The book is organized as a sequence of investigations. Each conversation centers on a concrete scientific question: What is gravity? What causes cosmological redshift? What is the wavefunction? What binds the nucleus? What would it mean to engineer coherence directly? In each case, Dr. S presents the strongest standard account. Dr. U presents the strongest UFD account. Then the real work begins: clarification, cross-examination, pressure-testing, and the search for decisive empirical leverage.

Readers should therefore expect neither a forced tie nor a predetermined triumph. There will be places where the standard framework appears stronger. There will be places where UFD appears stronger. There will also be places where the honest verdict is mixed. A serious inquiry must allow all of those outcomes.

The larger wager behind this project is simple. If reality is best described by a collection of highly successful but only partially connected theories, then this dialogue should help make that clearer. If, however, reality is better described by a smaller set of geometric and field-based principles that unify matter, light, interaction, cosmology, and perhaps even mind, then that too should become clearer as the conversation unfolds. In either case, the destination is the same: not loyalty to a paradigm, but truth. That is the spirit in which these conversations are offered.

What Counts as a Serious Alternative Theory?

Patchwork or Unity?

Modern physics is one of humanity's greatest intellectual achievements. It has measured, predicted, and organized an astonishing range of phenomena — from atomic spectra to black holes to the large-scale structure of the cosmos. It has given us Quantum Mechanics, General Relativity, the Standard Model of Particle Physics, and an understanding of the universe so precise that we can predict the outcomes of experiments to many decimal places.

Yet its deepest picture of reality remains divided. Gravity is described in one mathematical language, quantum theory in another, and no one has succeeded in unifying them despite a century of effort. Nuclear structure is described through yet another framework, and cosmology still relies on constructs like dark matter and dark energy whose physical nature remains genuinely unknown. None of this erases the success of the standard framework. But it does raise a harder question: when should a theory that challenges such a framework be taken seriously?

Unified Field Dynamics (UFD) begins from a different possibility. It proposes that matter, light, forces, quantum behavior, and cosmological structure may all be different expressions of one deeper physical architecture — a hierarchy of nested fields in which stable patterns emerge from a common underlying medium. If that proposal is correct, then the right question is not whether UFD sounds unfamiliar. The right question is whether a framework of that ambition can be judged fairly, rigorously, and in advance by the same standards that apply to any serious alternative in physics.

Physics had not failed. That was the first thing Dr. S wanted made clear.

Its equations still launched spacecraft, predicted particle collisions, modeled the interiors of stars, explained semiconductors, and tracked the observable universe with astonishing precision. Any framework that proposed to replace part of that inheritance carried a severe burden. It was not enough to point to anomalies, unresolved questions, or aesthetic dissatisfaction with the current picture. One had to say, clearly and in advance, what standards a serious alternative would have to meet.

Dr. U agreed with that immediately.

Respect, he said, was the right place to begin. Not deference. Respect. And if the conversation was going to be worthwhile, the standards had to be shared.

That was where they began.

DR. S: Before we go further, I want one rule on the table. The burden of proof is not symmetrical. Standard physics already explains an extraordinary amount. A replacement framework does not earn seriousness just by promising a cleaner worldview.

DR. U: Agreed. But I want the companion rule on the table as well. A replacement framework should not be rejected merely because it changes the ontology of the theory it challenges.

DR. S: Meaning?

DR. U: Meaning the basic picture of what actually exists in the world. Every physical theory carries a picture of what is really out there — whether there are particles or fields, whether spacetime is fundamental or emergent, whether forces are primitive or derived from something deeper. That picture is the theory's ontology. If a new theory replaces spacetime curvature with pressure geometry, or point particles with stable field excitations, or the metric expansion of space with the evolution of a real physical medium, the first response cannot simply be, "But that isn't the standard ontology." Of course it isn't. That is what replacement means.

DR. S:: Fair. Then the standards have to sit deeper than vocabulary.

DR. U: Exactly.

Dr. S nodded. That was already progress. The first agreement was not about UFD at all. It was about how a new framework should be judged.

DR. S: Good. Then let us state the standards plainly. A serious theory has to cohere with itself before it claims to explain nature.

DR. U: Internal consistency first.

DR. S: Yes. Not just local consistency in one corner, but consistency across the domains where it claims to apply. If it uses one ontology for nuclear structure, another for cosmology, and a third for quantum theory, then it is not unified. It is patched together from pieces that happen to share a name.

Dr. U: Agreed.

DR. S: Second, it must account for what we actually observe. Not merely reinterpret observations in more satisfying language, but account for them.

DR. U: Agreed. A theory does not become scientific by sounding more physical. It becomes scientific by surviving contact with the world. Measurements, experiments, data — those are the tests that matter, and a theory either passes them or it does not.

DR. S: Third, it must be able to fail. This is the criterion Karl Popper introduced in the twentieth century as a way of distinguishing science from pseudoscience. A theory is scientific only if it makes predictions specific enough that they could in principle be shown to be wrong. A theory that explains everything, no matter what happens, explains nothing.

DR. U: Not just somewhere vague in the future. Cleanly, and in identifiable ways.

DR. S: Good. Because broad theories often hide behind vagueness. A serious one should say what would count against it.

DR. U: Especially if it claims deep unification. The stronger the shared ontology, the stronger the failure burden.

That mattered. They were no longer talking about attitude. They were building a standard.

DR. S: Fourth, explanatory reach. But let me say this carefully. Breadth alone is not a virtue. A theory does not become impressive just because it talks about everything.

DR. U: Of course. Reach only matters if the same principles are doing real work across domains.

DR. S: Exactly. If one architecture explains nuclear binding, quantum behavior, gravity, and cosmology through linked mechanisms rather than unrelated stories, that is a major theoretical virtue. If it merely repeats the same words everywhere, it is atmosphere.

DR. U: Then let us say it that way. Not scope alone. Linked explanatory scope.

DR. S: Good.

DR. U: And I would add this: explanatory reach matters especially when it reduces the number of separate, unconnected mechanisms a theory has to postulate. If several standard mechanisms can be replaced by one shared geometry, that counts for something.

DR. S: It counts only if the geometry constrains the cases it claims to unify. A unified theory has to earn its unity by making testable commitments, not just by sounding unified.

DR. U: Agreed.

They paused. That was one of the first real hinges. Neither man wanted a theory rewarded merely for ambition. Both agreed that unity had to be earned through real explanatory work before it became a virtue.

DR. S: Fifth, honest counting of free parameters.

DR. U: Yes. That one matters more than people admit.

DR. S: Let me explain what I mean. Every physical theory contains some number of quantities that have to be fixed by measurement rather than derived from the theory itself. The mass of the electron, the strength of the electromagnetic force, the cosmological constant — these are free parameters that the theory does not predict but has to absorb from experiment. A broad theory should not buy its reach by quietly multiplying these adjustable inputs. The question is not how many symbols appear on the page. The question is how many genuinely independent knobs remain after calibration, reuse, and structural constraint. A theory with a hundred free parameters can fit almost any data. A theory with five that still fits the same data has done something much deeper.

DR. U: Right. So the bookkeeping has to stay honest. What is postulated? What is fixed once? What is derived? What remains open?

DR. S: Good. Because without that distinction, every ambitious framework can make itself look deeper than it is.

DR. U: And I would add one more refinement. Not every claim in a developing framework has the same maturity.

DR. S: Go on.

DR. U: Some claims are fully derived. They follow rigorously from the framework's core assumptions and have been carried to definite quantitative closure. Some claims are constrained but not fully closed — the theory has identified a real mechanism and narrowed the space of acceptable explanations but has not yet carried the result all the way to a first-principles derivation. And some claims are programmatic: they identify

a plausible mechanism and a coherent research pathway, but the full
working-out is still pending.

DR. S: That distinction should absolutely be on the table from the start.

DR. U: It protects both sides. It prevents the framework from inflating its
claims on one side, and it prevents the critic from dismissing unfinished
work as if it were equivalent to failed work on the other.

DR. S: Good. Then that is one of our rules. A theory must not pretend
equal maturity everywhere.

That was important enough to linger. A derived claim was the strongest
kind. It meant that the framework had carried a result to real closure: the
mechanism was specified, the reasoning was explicit, and the outcome had
reached a definite structural or quantitative result rather than remaining a
suggestive sketch.

A constrained claim stood one level below that. It meant the theory had
identified a real mechanism and narrowed the space of acceptable
explanations, but had not yet carried the result all the way to full first-
principles closure. Something important had been gained, but the final
derivation remained incomplete.

A programmatic claim was weaker still, though not empty. It meant the
framework could point to a plausible mechanism, a coherent research
direction, and a set of future burdens, but had not yet earned the right to
speak as if the result were established. It was a map of where the theory
thought the physics should go next, not a destination already reached.

A weaker theory would have wanted every claim to sound equally finished.
A weaker critic would have wanted every unfinished part to count as equal
failure. Neither move would be honest.

So the standard became sharper.

DR. S: Sixth, mathematical rigor.

DR. U: Naturally.

DR. S: By which I do not mean impressive-looking equations that do not
actually derive anything. I mean: do the derivations actually follow from the
assumptions? Are approximations marked? Are open terms admitted as
open? Is the mathematics adequate to the claim it is supposed to support?

DR. U: Yes. The burden is not "does the math look impressive?" It is "does
the framework know what it has actually established?"

DR. S: Good. Because a theory can overstate itself in equations just as easily as in prose.

DR. U: Sometimes more easily.

Dr. S smiled faintly at that. Then he added something else. Not a new criterion exactly. A methodological warning.

DR. S: I want one more rule. Direct empirical contact should carry more weight than constraints inherited from the theory being replaced.

DR. U: Meaning?

DR. S: Suppose a physicist calculates a bound on some quantity — a maximum possible mass, a minimum possible temperature, a limit on how strongly two things can interact. That bound is derived using the assumptions of the current framework. It depends on the current ontology. If a new theory proposes a different ontology, then the bound may no longer apply in the same way. It may still matter. But it does not automatically transfer as a refutation because the reasoning that produced it was internal to the theory being replaced.

DR. U: What transfers cleanly are the things we actually measure.

DR. S: Exactly. Spectra, timing relations, morphology, scaling laws, transport signatures. Those are observations, not interpretations. If two theories explain the same observation through different mechanisms, then one theory's intermediate calculations are not automatically the other theory's burden.

DR. U: Good. Then I'll say it plainly. A replacement theory is judged first by what it actually predicts about observations, by whether its derivations follow on its own terms, and by whether it can fail cleanly — not by how faithfully it preserves the chains of reasoning of the theory it seeks to replace.

DR. S: Yes. That is the right formulation.

That changed the conversation. Because now the standard was not only demanding. It was fair. A serious alternative would not be protected from data. But neither would it be strangled in advance by assumptions it explicitly denied. That was exactly the balance the conversation needed.

DR. S: I want one final methodological point. A unified framework should be judged both locally and globally.

DR. U: Meaning?

DR. S: Meaning one can ask whether each sector of the theory stands up on its own terms. Does the nuclear physics work? Does the cosmology work? Does the quantum mechanics work? But one must also ask whether the framework remains coherent when read as a whole. Does the nuclear physics fit with the cosmology? Do the quantum mechanics and the relativity speak the same ontological language?

DR. U: Yes. Because the central claim of a unified theory is not that it contains several interesting sectors. It is that the same principles survive export across domains.

DR. S: Exactly. Local rigor matters. Holistic coherence matters too.

DR. U: And if the same principle replaces several otherwise independent mechanisms across scales, then success there should count more than isolated local success.

DR. S: It should. But the reverse is also true. If the framework is genuinely unified, then foundational failure should propagate. A problem in one sector should affect the others, because the sectors share their foundations.

DR. U: Agreed. A deep theory should fail along its joints.

That was the strongest agreement yet. A weaker theory fails in compartments — one part breaks while the rest carries on as if nothing had happened. A stronger theory makes itself vulnerable at the places where its principles connect. Dr. S liked that. Dr. U liked it too. Which was one reason the conversation had become more serious than either had expected.

DR. S: Let me see if I can summarize the standard we are building.

DR. U: Go ahead.

DR. S: A serious replacement framework must satisfy at least six core criteria: internal consistency, accounting for what we observe, the ability to fail cleanly, linked explanatory reach, honest counting of free parameters, and mathematical rigor that matches its claims.

Dr. U: Yes.

DR. S: It must also distinguish the maturity of its claims: what is derived, what is constrained, and what remains programmatic.

DR. U: Yes.

DR. S: It should be judged both locally and holistically — sector by sector and as a whole.

DR. U: Yes.

DR. S: It should not be penalized merely for changing ontology.

DR. U: Correct.

DR. S: But it should be judged by what it actually predicts about observations, by whether its derivations follow on its own terms, and by whether it can fail cleanly.

DR. U: Exactly.

DR. S: And because it is a replacement framework, its real burden is not simply to disagree with the standard model of physics. It is to explain more with fewer independent assumptions while remaining fully exposed to correction.

DR. U: That is the right standard.

They sat with that for a moment. Because the standard had become real enough now to be uncomfortable. And that was the point. A good standard should not flatter the theory being judged. It should threaten it. This one did.

DR. S: Good. Then there is one last thing I want to add.

DR. U: Which is?

DR. S: We should not confuse empirical strength with ontological completion. Standard physics may still outperform a challenger in precision, maturity, and institutional depth across many domains. That matters. But it does not settle whether its ontology is final. A theory can predict everything and still leave its deepest questions unanswered.

DR. U: And the inverse is also true. A replacement framework may offer a more coherent picture of what the world is while still lacking full closure in some sectors. That coherence does not excuse the missing closure.

DR. S: Good. Then we agree on that too. A theory can be extraordinarily good at predicting observations and still leave unanswered what it is really talking about. Or it can offer a deeper picture of what the world is and still fall short when it tries to predict specific numbers.

Dr. U: Exactly.

DR. S: Which means the real question is never just "does it work?" or "does it sound deeper?" It is whether it meets the standards we have just named more successfully, more honestly, and more integratively than the framework it challenges.

DR. U: That is right.

That was the deepest agreement of the opening exchange. Not that standard physics was finished. Not that UFD was right. But that both would be judged by a common standard that neither could evade.

That was enough. Not agreement on the world. Agreement on how the world should be argued about. And perhaps that was the only kind of beginning worth trusting.

DR. S: Let me try the summary test.

DR. U: Go ahead.

DR. S: A serious replacement framework is not judged by novelty, by atmosphere, or by whether it preserves inherited ontology. It is judged by whether it coheres with itself, survives empirical contact, exposes itself to failure, explains more with one linked structure, disciplines its assumptions, and knows what its mathematics actually establishes. It must distinguish finished derivations from constrained but open claims, and those from genuine research pathways. It must stand up both sector by sector and as a whole. And because it aims to replace rather than extend, it must be tested first against direct observables and explicit failure conditions, not against every intermediate construct of the paradigm it disputes.

DR. U: That's fair.

DR. S: Good.

DR. U: Good?

DR. S: Yes. Because now the burden is no longer vague.

DR. U: It shouldn't be.

DR. S: And neither of us gets to move the goalposts later.

DR. U: That may be the most important agreement we've reached.

INTERLUDE

By the end of the first exchange, neither man had defended a theory yet.

That changed the temperature.

They had done something more important first: they had agreed on the standards by which any defense would later stand or fall.

Those standards were demanding enough to threaten both sides. Standard physics retained the burden of explaining why its extraordinary success should count as more than precision inside a structurally fragmented inheritance. UFD retained the heavier burden of showing that a replacement ontology could do more than sound coherent. It would have to survive empirical contact, parameter discipline, explicit failure risk, and the pressure of being judged as one framework rather than as a set of isolated local stories.

That was a good beginning.

A theory should not be protected by sympathy.
A critique should not be protected by inherited assumptions.
The argument should become harder for both sides as soon as it becomes fair.

Now it had.

TEMPORARY VERDICT

Dr. S succeeded in forcing the essential discipline. A replacement theory does not earn attention simply by challenging a successful framework or by offering a more satisfying picture. It must submit to hard scientific standards and remain vulnerable where it claims the most.

Dr. U succeeded in forcing the equally important correction. A replacement theory should not be judged by whether it preserves the ontology of the paradigm it seeks to replace, but by whether it offers a more coherent, more testable, and more integrative account of reality on its own terms.

So the conversation begins where it should.

Not with agreement about the world.

With agreement about the standard by which both pictures must answer for it.

CHAPTER 1

WHAT IS UFD, REALLY?

A WORLD OF PARTICLES, OR A WORLD OF PATTERNS?

Modern physics tells a story about particles and fields. The particles are the main characters — electrons, quarks, photons, and the rest of the small inhabitants of the microscopic world. The fields are what let them act on one another. An electron in one place can push another electron across the room because the electromagnetic field between them carries the push. Quarks cling together inside a proton because the gluon field between them holds them there. In the standard telling, particles are real things and fields are what connects them.

The resulting system works extraordinarily well. It predicts, classifies, and calculates with astonishing success. The Standard Model of particle physics has been confirmed to remarkable precision. Quantum electrodynamics, the theory of how electrons and photons interact, makes predictions that match experiment to more decimal places than almost any other scientific claim in history. The picture of particles-as-primary has earned its place through a century of empirical success.

And yet the overall picture can still feel divided. Gravitation is described one way, through the curvature of spacetime in Einstein's General Relativity. Quantum behavior is described another way, through wave functions evolving in abstract mathematical spaces. Nuclear binding is described through yet another framework, the residual strong force between quarks. Cosmology adds still further structure, with dark matter and dark energy standing in for phenomena the standard framework cannot otherwise explain.

Further still, fields themselves are treated as mathematical structures — assignments of values to points in abstract spaces — with the question of whether they correspond to real physical media usually set aside as philosophical rather than scientific. The achievement is undeniable. The physical picture is less fully unified, and less "physical," than the success of the mathematics sometimes makes it seem.

Unified Field Dynamics, or UFD, begins from a different starting point. It treats fields as primary and stable objects as what fields do under the right conditions. Matter is not built first from tiny independent particles moving through emptiness. It is built from stable patterns — vortices, standing waves, and organized structures that hold their shape — sustained within real physical media referred to in the literature as a *plenum*. An electron, on this view, is not a tiny point of stuff that happens to interact with a field. It is a particular kind of stable pattern in the field itself. A proton is a different kind of stable pattern. A photon is a wave-like excitation that propagates through the field. The field is what is really there. The particles are what the field looks like when it organizes itself into persistent forms.

In its full form, UFD is organized as a hierarchy of four nested fields, each with its own physical role and characteristic kind of stable structure. At the deepest level is the Universal String Field, or USF — the vibrating constraint layer that governs which forms, topologies, and degrees of organization are even possible. The USF does not itself carry mass or propagate signals; it sets the rules for what the higher layers can do. From the USF emerges the Universal Energetic Field, or UEF — the dense energetic medium associated with mass, nuclear binding, and the pressure geometries that UFD proposes in place of gravitational curvature. The UEF is the layer where heavy, bound structures live. Above the UEF lies the Universal Light Field, or ULF — the lighter propagation field associated with light, electromagnetism, and quantum mechanics. The ULF is the layer where waves travel and where the quantum behavior of ordinary matter takes place. And in the fuller downstream picture, there is a still more weakly coupled layer, the Universal Coordination Field, or UCF, tied first to the neutrino sector and later, in the book's most ambitious extension, to consciousness and subjectivity itself.

The point of the hierarchy is not to multiply invisible entities for their own sake. It is to explain why different kinds of stable structure appear at different levels of reality: why some forms are heavy and bound, why others are light and propagating, and why the organized world may reflect a descent through progressively simpler and more weakly coupled field layers. Each layer constrains the layers above it. Each layer gives rise to the layers below it. The whole architecture is meant to be one continuous physical reality, differentiated into distinct regimes rather than divided into unrelated compartments.

The world, on this view, is less like a box of separate things and more like a layered physical continuum capable of sustaining persistent forms.

At this point in the conversation, Dr. S no longer needed the glossary. He knew the outline. Real fields. A nested hierarchy. Stable excitations rather than primitive particles. A cascade rather than a stack. He knew enough, in other words, to become dangerous. That was useful.

The first exchange had been about standards. This one would be about intelligibility. Not whether UFD had ambition. It plainly did. The question now was whether it described a world one could actually believe in.

What was at stake in this conversation was not yet whether UFD was true, but whether it offered a genuinely different picture of reality. Most readers are used to hearing that physics is about particles moving under laws. UFD begins from a different intuition: fields first, stable objects second. The question of the chapter was whether that intuition could be turned into something more than a suggestive rearrangement of vocabulary.

Dr. S began with the impatience of someone willing to tolerate large ideas, but only if they survived sharp questions.

Dr. S: All right. I understand the outline. Real fields, layered structure, particles as stable patterns rather than fundamental building blocks. Fine. My question is simpler than that. Why should I think this is more than a prettier vocabulary for things we already know how to calculate?

This was the right opening question. A new picture of what exists in the world earns nothing if it merely renames familiar results in more attractive language. The issue was not whether UFD sounded deeper, but whether it changed what kind of thing reality was claimed to be.

Dr. U: Because it changes what kind of thing the world is.

Dr. S: That is exactly the sort of sentence that makes physicists nervous.

Dr. U: It should. But it is still the right sentence. Standard physics often starts with particles as the basic furniture of the world and fields as the structures that govern or connect them. UFD reverses that. Fields are basic. Stable objects are what fields do when the geometry and organization are right.

Dr. S: So electrons, protons, atoms—

Dr. U: Stable organizations of underlying media.

DR. S: Media in the literal sense?

DR. U: Yes. Not a metaphor. Real continuous substances capable of supporting persistent forms.

DR. S: Then let me stop you there. If the word "field" is doing this much work, I want it defined before we go any further.

DR. U: Fair. In UFD, a field is not just a mathematical assignment of values to points. It is a real physical medium — something extended, continuous, and capable of bearing pattern, stress, propagation, and stable form.

DR. S: So when you say "field," you mean something more like a plenum than a bookkeeping device.

DR. U: Exactly. A plenum — a fully filled medium with no true empty space. Not empty space decorated with numbers, but an actual substance with physical properties.

DR. S: All right. Then define energy too, while you are at it. Because if the world is field-first, energy cannot just remain a floating abstraction.

DR. U: It cannot. In UFD, energy is what the field is doing — how it moves, circulates, stretches, loads, or holds a stable form. More precisely, energy is the motion and geometric configuration of a real field.

DR. S: So energy is not something poured into the field from outside.

DR. U: No. It is a state of the field itself.

DR. S: That helps. A great many theories talk as though energy were an invisible fluid added to matter after the fact. You pour energy into a system, and things happen. Energy flows out of a system, and things slow down.

DR. U: That picture has done useful work in practice, but it carries a philosophical cost. It makes energy sound like a separate substance that keeps company with matter rather than a feature of matter's own activity. In UFD, that split disappears. The field is the real medium. Energy is its activity and configuration. Matter is a stable organization within it.

DR. S: Which means the basic picture is no longer "things in space," but "patterns in a plenum."

DR. U: Exactly.

Dr. S paused.

DR. S: That is a clean idea. It is also an old temptation. Physics has repeatedly tried to make the world feel more physically imaginable by positing some underlying medium. The most famous attempt, the luminiferous ether of the nineteenth century, was supposed to be the medium through which light waves propagated. It was eventually abandoned when the Michelson-Morley experiment of 1887 failed to detect any motion of the Earth through it, and Einstein's special relativity of 1905 made the ether unnecessary as an explanatory device. Usually, when physics posits a hidden medium, it goes badly.

DR. U: Usually because the medium is introduced after the fact, beneath a framework that was never built for it. The ether was added to classical physics as an afterthought — a passive background substance whose only job was to carry light waves. When it turned out to leave no detectable traces, the framework that had been built without it had no trouble throwing it away. UFD does not work that way. It does not say, "Here is standard physics, and beneath it let us tuck a hidden fluid." It says the world is field-first from the beginning. The stable things come later.

DR. S: So the ontology is not a retrofit.

DR. U: No. It is the starting point.

That answer helped. Not because it settled anything, but because it made the disagreement cleaner. Dr. S was not yet asking for equations. He was asking whether the picture of reality had necessity.

DR. S: Fine. But I still want to know why you need more than one field. If your instinct is unity, why not one universal field doing everything?

This was an important pressure point. A theory that promises deeper unity can easily undermine itself by multiplying invisible layers whenever needed. Dr. S was now asking whether the four-field hierarchy was a necessity or just a new way of organizing complexity.

Dr. U: Because undifferentiated unity explains less than people think. A world can be one in its source and still differentiated in its functions.

DR. S: That sounds suspiciously philosophical.

DR. U: It is philosophical. We are talking about what the world is actually made of at the deepest level.

DR. S: Fair enough. Then why the differentiation?

DR. U: Because constraint, substance, and propagation are not the same physical role. A layer that sets rules about what kinds of stable form are even possible is not yet the same as a dense medium that supports binding and mass. And neither of those is the same as a lighter field that carries wave propagation, electromagnetic structure, and quantum coherence — the organized phase relationships that make quantum behavior possible. Each of those roles is physically distinct. A single undifferentiated substance would have to do all of them at once, and there is no reason to think a single substance could play such different roles equally well.

DR. S: So the hierarchy is doing conceptual work before it does quantitative work.

DR. U: Right. The mistake is to think that "one world" must mean "one undifferentiated stuff." It may instead mean one world with lawful descent into distinct regimes.

DR. S: Descent. That is the key word for you, isn't it?

DR. U: It is. The hierarchy is not a filing cabinet. It is a cascade.

DR. S: Explain that without preaching.

Dr. U smiled.

DR. U: Think of the way water flows from a high reservoir through a sequence of progressively smaller channels — first through a dam, then down a river, then into smaller streams, then into tributaries, then into the fine capillary branches of a delta. Each stage is still water. Each stage has its own characteristic behavior: the reservoir holds still under pressure, the dam channels flow through narrow outlets, the river carries large volumes slowly, the streams carry smaller volumes more quickly, the capillaries spread in fine branches. The whole system is connected, and what happens at each stage is shaped by what comes before it. That is the kind of structure the UFD hierarchy is meant to describe. The deepest layer, the USF, is the reservoir — the layer that constrains what kinds of stable form are even possible. It is not yet the dense, substantial medium in which binding and massive structure occur. So there is a transition into a more substantial regime, the UEF, which is where the heavy bound structures live: nuclei, mass, the pressure geometries that replace gravitational curvature. Then that regime yields a lighter interaction field, the ULF, which carries light, electromagnetism, and quantum behavior. In the fuller framework, the hierarchy may continue into a still weaker layer, the UCF, which is associated first with neutrinos and later with the question of consciousness itself. The layers are related by descent and by what each layer has to bear, not by arbitrary multiplication.

DR. S: So not "four unrelated things," but one reality differentiating into several physically distinct regimes.

DR. U: Yes. That is much closer.

Dr. S tapped his pen lightly.

DR. S: I can see the appeal. It takes the patchwork of physics and turns it into a genealogy.

DR. U: Exactly.

DR. S: But genealogy can turn into mythology very quickly.

DR. U: Only if it stops constraining anything.

Dr. S: Good. Then let us press there. What does the hierarchy actually buy you?

DR. U: A different vocabulary for describing how things happen. Once fields are primary, you stop reaching first for fundamental particles and separately postulated forces. You start asking about coherence — how well different parts of a system hold together as one organized whole. You ask about gradients — how a quantity changes from one place to another. You ask about topology — the mathematics of shape and connection, which studies which forms can be smoothly transformed into each other and which cannot. You ask about boundaries, about phase stability, about resonance, and about loading.

DR. S: Loading?

DR. U: Different field layers do not carry the same kind of physical burden. Think of the way a suspension bridge bears load differently from a trampoline. A suspension bridge holds its shape under enormous weight because its structure is built to distribute and transmit that weight to anchored points. A trampoline responds to weight by deforming elastically and returning to its original shape. Both are real surfaces capable of bearing load, but the kind of load they bear and the way they respond to it are fundamentally different. A dense binding field is not loaded like a light propagation field. Their characteristic responses should therefore differ in lawful ways.

Dr. S: And you think the constants of physics are traces of that difference.

Dr. U: Some of them, yes. That is the deeper ambition.

Dr. S: Ambition or result?

Dr. U: Both, depending on the case. The careful version is this: UFD does not need every physical constant — every fixed number like the speed of light, the charge of the electron, or the strength of gravity, which standard physics measures but does not derive from anything deeper — to be fully explained from first principles on day one. But it does need those constants to stop looking flat. They should begin to look like traces of hierarchy, of how stiff each layer is, of how things circulate within it, and of the topological shapes that the layers can support, rather than a disconnected list of measured gifts.

That changed the feel of the conversation. The hierarchy was no longer just a way of classifying fields. It was starting to behave like an architecture with consequences.

Dr. S pushed again.

Dr. S: All right. Suppose I grant, provisionally, that fields are primary and the hierarchy is real. What actually changes in the way you read physics?

Dr. U: Almost everything.

Dr. S: That had better not be rhetoric.

Dr. U: It is not. Gravity becomes pressure geometry in a dense field rather than a primitive curvature of spacetime. Inertia — the resistance of matter to being accelerated — becomes the impedance of a lighter field, the way a medium resists being disturbed, rather than a brute property that matter somehow has on its own. Quantum states become real resonant structures in the field rather than merely formal mathematical amplitudes. Nuclear binding becomes coherence and the geometric way that nuclear components lock together, rather than a primitive exchange force at the deepest level. Cosmological redshift — the reddening of light from distant galaxies — becomes the behavior of a real propagating medium rather than a metric expansion of space as the only imaginable story.

Dr. S: So this is not a local reinterpretation of one puzzle.

DR. U: Not remotely. It is a different physical picture of what the world is doing.

DR. S: Which is exactly why the burden is so high.

DR. U: Yes. A framework like this earns freedom at the level of ontology only by becoming more accountable everywhere else.

That mattered. A less serious approach would have wanted the comfort of reinterpretation without the cost of replacement. A stronger one accepted that changing the picture of reality meant taking on a much wider burden.

Dr. S then moved to the worry he had been carrying from the start.

DR. S: Let me tell you what still bothers me. This all hangs together because the words are compatible: field, coherence, topology, descent, loading, resonance. Fine. But how do I know this is a physical framework and not just a very elegant vocabulary net — a collection of terms that sound related but do not actually restrict what can happen?

DR. U: Because the concepts are supposed to constrain, not just harmonize the tone.

DR. S: Easy to say.

DR. U: Then I'll say it more sharply. If fields are real media, some kinds of behavior should become impossible and others natural. If particles are stable excitations rather than fundamental things, their identities should follow from what the fields can support — you should not be free to postulate any particle you like, because the fields will only support certain kinds of stable structure. If the hierarchy is real, the constants of physics and the rules that connect them should begin to look related rather than independent. If coherence is fundamental, then binding, propagation, and breakdown should repeatedly reduce to questions about how organized phase relationships are maintained or lost across scales.

Dr. S: So the framework lives or dies by whether the concepts become restrictive.

DR. U: Exactly. If they do not reduce freedom, they are decoration.

That was the center of the exchange. Not whether UFD was imaginative. Whether it was restrictive enough to deserve the name of a serious physical framework.

Dr. S heard that clearly and softened, though only slightly.

DR. S: All right. Let me be fair. There is something genuinely attractive here. Standard physics can feel like an inventory of local triumphs — a list of separate successes held together mainly by the fact that they all use mathematics. This gives you one picture in which the triumphs are supposed to become relatives rather than neighbors.

DR. U: Yes.

DR. S: The danger, of course, is that relatives can be invented after the fact.

DR. U: Of course.

DR. S: So the real question is whether the family resemblance is discovered or retrofitted.

DR. U: Exactly.

DR. S: And what is the standard picture missing, in your view?

DR. U: A reason why the pieces belong together at all.

DR. S: That sounds too easy. Standard physics knows very well that its constants and sectors are puzzling. Physicists have been trying for decades to understand why the fundamental constants have the values they do, and why the Standard Model has the particular structure it has. They are not unaware of these questions.

DR. U: Of course. I am not saying standard physics is careless. I am saying it is methodologically conservative. It treats the constants as measured anchors and moves on. That is often the right move for calculation, and it has produced most of modern physics. But it leaves open the deeper question of whether those anchors are structurally related — whether the reason the speed of light is what it is, and the reason the electron has the mass it has, and the reason gravity is as weak as it is, might all be connected by a single underlying architecture.

DR. S: And UFD wants to treat them more like fossils than like arbitrary pegs.

DR. U: Exactly. Fossils of the descent structure. Just as a paleontologist reads the fossils in a cliff face to reconstruct the history of life, UFD wants to read the constants of physics as traces of how the field hierarchy unfolded — signatures that something specific happened, and that the specific thing still shapes what we see.

Dr. S: That is a good phrase.

DR. U: It is also the burden. Because once you say that, the architecture had better leave real traces.

That line stayed with them a moment. The conversation was no longer about whether UFD was more intuitive. It had become a question about whether reality's visible constants and stable forms might already be evidence of a layered architecture rather than brute givens.

Dr. S then pushed where the chapter naturally wanted to go next. Mass ratios. Scaling. Constants. The bridge to specific quantitative predictions had appeared.

DR. S: Let me try the summary test.

DR. U: Go ahead.

DR. S: On your view, UFD is not just standard physics with nicer metaphors. It is a field-first picture of reality. Stable objects are not fundamental building blocks but persistent patterns in real media. A field is not merely a mathematical overlay on empty space but an actual physical substance, and energy is the motion and configuration of that substance rather than an independent abstraction poured into it from outside. The hierarchy of fields is meant to do real physical work: one layer constrains what kinds of stable form are possible, another supports binding and the pressure geometries that replace gravitational curvature, another supports light and quantum coherence, and the fuller framework proposes a further coordination layer whose status must be earned rather than assumed. The ambition is not merely to relabel known phenomena, but to show that what standard physics treats as partly separate sectors may be related by one underlying architecture. And if that architecture is real, the constants and hierarchies of physics should begin to look less like a flat list of measured inputs and more like traces of descent, loading, and topological shape.

DR. U: That is fair.

DR. S: The attraction is obvious.

DR. U: Yes.

DR. S: The danger is equally obvious.

DR. U: Yes.

DR. S: Good. Then the burden is finally clear.

INTERLUDE

By the end of the second exchange, the issue was no longer whether UFD offered a more satisfying picture. It had become whether that picture was restrictive enough to deserve belief.

That mattered.

Many ambitious frameworks fail not because they lack imagination, but because their unifying language never becomes costly. It never forbids enough. It never narrows enough. It never forces real relations among things that had previously appeared separate.

That was now the pressure on UFD.

If reality is indeed built from nested fields, stable excitations, and coherence structures, then the hierarchy cannot remain a pleasing conceptual scheme. It must leave visible traces: in constants, in scaling relations, in mode structure, in transformation rules, and in the way domains now treated as separate begin to fit together.

The attraction remained.

So did the burden.

And that was exactly as it should be.

TEMPORARY VERDICT

Dr. U succeeded in making UFD intelligible as a genuine ontology rather than a decorative reinterpretation. Reality became field-first, particles became stable excitations, energy became the activity and configuration of real media, and the hierarchy became a cascade of physically distinct regimes rather than a pile of separate substances.

Dr. S succeeded in forcing the decisive pressure. A satisfying picture of the world is not enough. The concepts must become restrictive. They must reduce arbitrariness, not merely redescribe it.

So the conversation ends where it should.

IF, HOWEVER, UFD IS REAL, THE CONSTANTS SHOULD STOP LOOKING LIKE A DISCONNECTED INVENTORY AND START LOOKING LIKE STRUCTURAL CLUES. THAT IS THE NEXT BURDEN TO BE ADDRESSED.

CHAPTER 2

TOPOLOGICAL DESCENT

FROM TREFOIL TO TORUS TO SPHERE

The twentieth century turned the question of what the world is made of into one of the great scientific stories. At the beginning of the century, atoms were still widely treated as the fundamental building blocks of matter — the Greek word *atomos* means "uncuttable," and for most of the history of chemistry that is exactly what atoms were taken to be. Then, in 1897, J. J. Thomson discovered the electron, the first clear indication that atoms had internal parts. In 1911, Ernest Rutherford showed that atoms have tiny dense nuclei at their centers, with electrons surrounding them. In 1932, James Chadwick identified the neutron, completing the basic picture of the atomic nucleus as a collection of protons and neutrons. By the middle of the century, physicists thought they had a manageable list: protons, neutrons, electrons, photons, and the newly discovered neutrino.

Then things got stranger. As particle accelerators grew more powerful in the 1950s and 1960s, they began producing a flood of new particles that nobody had expected and nobody could easily explain. Pions, kaons, muons, hyperons, and resonances of every description multiplied rapidly. The question of what the world was made of had become a question of how to make sense of a zoo.

The Standard Model of particle physics, developed through the 1960s and 1970s, brought order to that zoo. It showed that most of the observed particles were not truly fundamental but were built from a smaller number of deeper components called quarks. Protons and neutrons are each made of three quarks bound by the strong interaction. Lighter particles like pions and kaons are made of quark-antiquark pairs. Electrons, muons, and neutrinos belong to a separate family called leptons. A handful of force-carrying particles — photons, gluons, and the W and Z bosons — mediate the interactions between them. The resulting framework predicts particle behavior with extraordinary precision. It is one of the most successful scientific theories ever constructed.

And yet the picture can still feel, at a physical level, more like a brilliantly ordered inventory than like a transparently unified hierarchy. The Standard

Model tells you what the particles are, what their masses are, how they decay, and how they interact. It is less forthcoming about why there are exactly these particles, why their masses take the values they do, why the coupling strengths fall where they do, or why the spectrum has this particular architecture rather than another. The mathematics is extraordinarily successful. The deeper question of why the catalog has the contents it has is usually left open.

Unified Field Dynamics starts from a different physical picture. It treats particles not as tiny independent objects dropped into the world, but as stable patterns that different layers of a real field hierarchy can sustain. In UFD, the proton belongs to the Universal Energetic Field (UEF), the electron to the Universal Light Field (ULF), and the neutrino to the Universal Coordination Field (UCF). Different layers of the hierarchy support different kinds of durable form, much as different physical media can support different kinds of waves, vortices, or stable structures.

A rope can hold a knot that water cannot. Water can hold a whirlpool that air can only briefly sustain. Air can carry a sound wave through distances that neither a rope nor a whirlpool would survive. Each medium has its own characteristic kinds of stable form, determined by the medium's own properties. In UFD, the same logic applies to the field hierarchy. A particle is not a tiny object inserted into a field. It is a stable pattern the field can hold.

The denser and more heavily loaded the field, the more intricate the pattern it can sustain. The lighter and more weakly coupled the field, the simpler the stable structures become. Matter therefore descends not only in density and coupling, but also in topology — in the deep structural type of pattern each field layer can support.

In this setting, topology is the mathematics of shape and connection. Two shapes have the same topology if one can be continuously deformed into the other without cutting or gluing. A coffee cup and a doughnut have the same topology because both have one hole through them; a sphere and a doughnut do not, because a sphere has no hole. A knot tied in a piece of rope has a different topology from an unknotted loop, because no amount of stretching or bending can turn one into the other without cutting the rope. In UFD, these distinctions are not just mathematical curiosities. They correspond to real differences in what kinds of stable structure a field layer can support.

The claim is not merely that particles have shapes. It is that particle identity depends, at least in part, on which kinds of stable topological structure a given field layer can actually sustain. The proton, electron, and neutrino are not just particles with different masses and charges. They are different kinds of coherent structure living in different layers of the hierarchy. A particle remains the kind of thing it is because the field has settled into a durable pattern, and that pattern is protected by its topology. It may interact, deform, and in some cases decay, but it does not casually turn into an entirely different kind of structure, because that would require a change in topology itself.

More precisely, UFD ties particle identity to a stable or metastable topological class in the underlying field. A fully stable structure, like the electron, does not decay as far as anyone has ever measured. A metastable structure, like the neutron, is protected only conditionally and can eventually unravel into simpler forms under the right conditions. UFD reads that difference topologically: some structures are fully protected, while others are only temporarily so.

That is what gives the hierarchy its explanatory force. The proton, electron, and neutrino are not assigned to different layers because the framework happens to contain multiple fields. They occupy different layers because those layers support different kinds of coherent structure. The proton is a knot-like excitation in the dense Universal Energetic Field. The electron is a toroidal vortex in the lighter Universal Light Field. The neutrino is a closed, loopless, knotless excitation in the still weaker-coupled Universal Coordination Field. Each lives where it does because that is where its kind of structure is possible.

The burden of the theory is to show that the stability, decay pathways, mass ratios, and interaction signatures of these particles all follow from the same underlying geometry. If UFD is right, then the particle spectrum should not look like a flat inventory that physicists happen to have discovered. It should look like evidence of a hierarchy.

By this point in the conversation, the vocabulary of the framework had begun to settle. Fields. Descent. Loading. Coherence. Geometry. But one question still hung in the air with unusual force. If UFD was right that the world was built from nested field regimes rather than from one flat inventory of particles, then the particle spectrum itself ought to show that hierarchy. It should not look random. It should look ordered.

Dr. S put the problem plainly.

DR. S: All right. I understand the fields well enough to be worried about them. Now I want to know whether the particle story is actually cleaner than the standard one, or whether you are just replacing one table of entities with another. Why these particles? Why this hierarchy? Why proton, electron, neutrino in this order?

That was the right pressure point.
A hierarchy only earns its place if it explains why the spectrum looks ordered rather than arbitrary. Otherwise, UFD would merely be replacing one particle table with another, only with more suggestive language.

DR. U: Because the hierarchy is not just a hierarchy of fields. It is a hierarchy of stable patterns those fields can sustain.

DR. S: Go on.

DR. U: Different field layers can hold different kinds of durable structure. Think back to the image from earlier: a rope can hold a knot, water can hold a whirlpool, air can carry a sound wave. Each medium has its own stable repertoire. UFD says the field hierarchy works the same way. The denser and more heavily loaded the field, the more intricate the pattern it can stably sustain. The lighter and more weakly coupled the field, the simpler the stable structures become. So as coherence descends through the hierarchy, the patterns simplify. In the technical language of the theory, the topology simplifies with them.

DR. S: Hence topological descent.

DR. U: Exactly.

DR. S: And you mean that literally.

DR. U: Completely literally.

That was the first hard line of the discussion.
Not just that particles lived in different fields.
That much was already familiar.
The sharper claim was that particle identity itself reflected what topology each field could bear.

DR. S: Fine. Then make it concrete. Start with the proton.

DR. U: In UFD, the proton is the ground-state heavy structure — the lowest-strain stable configuration the dense Universal Energetic Field can sustain. It is a trefoil-like coherent vortex in that field. Not light matter made heavy by an arbitrary parameter. A tightly loaded, knotted structure in the field layer that can actually carry that burden.

DR. S: Why a trefoil, though? Why not some other knot? If topology is doing real work here, the choice cannot just be decorative.

DR. U: It cannot be decorative. The trefoil matters for two reasons at once. First, it is the simplest nontrivial stable knot — the most basic closed knot that cannot be smoothly undone. Second, it carries a natural threefold internal organization.

DR. S: And you take that to matter physically.

DR. U: Very much so. One of the strongest clues about the proton is that, when you probe it hard enough, it does not behave like a featureless lump. It behaves like a structured object with a robust threefold internal pattern.

DR. S: You mean the evidence that led to quark language.

DR. U: Exactly. In the late 1960s, high-energy scattering experiments showed that the proton did not respond like a smooth ball. It responded as though it contained three distinct momentum-bearing centers. In standard language, that became the empirical basis of the quark model. UFD does not deny any of that evidence. It asks what kind of underlying field structure would naturally present that signature.

DR. S: And your answer is: a three-crossing knot.

DR. U: Exactly. The point is not merely that the proton is knotted. It is that the simplest stable knot already carries the kind of threefold internal organization the proton seems to reveal under high-resolution probing.

DR. S: So the trefoil is not just visually suggestive. It is your way of explaining why the proton looks three-partite under stress.

DR. U: That is exactly the point. In UFD, the crossings are not little beads hidden inside the proton. They are the geometric reason one coherent object can present three dynamically distinct internal sectors when it is probed hard enough.

DR. S: Which means you are not discarding the evidence that gave rise to quark language. You are reinterpreting it.

DR. U: Yes. The empirical burden remains the same. The question is whether the proton's threefold structure is better understood as evidence of three fundamental pointlike constituents, or as the projection of one stable knotted field structure with three internally distinct momentum-bearing regions.

DR. S: So heavy matter begins knotted.

DR. U: Yes.

DR. S: And the electron?

DR. U: The next stable rung down. A toroidal excitation — a stable doughnut-shaped vortex — in the lighter Universal Light Field. It keeps a loop, which means it still supports persistent circulation around an axis. That circulation gives the electron its charge, its magnetic structure, and its coupling to the electromagnetic sector. But it no longer carries the knotted burden of heavy matter. The topology is simpler: a single loop rather than a three-crossing knot.

DR. S: So the hierarchy is not just heavy versus light. It is knot versus loop.

DR. U: That is a good first approximation.

DR. S: Proton: trefoil. Electron: torus. What, then, is the neutrino?

DR. U: The endpoint of descent.

DR. S: Meaning?

DR. U: A closed, loopless, knotless excitation in the weakest-coupled layer — the Universal Coordination Field. In topological language, it is a genus-zero structure.

DR. S: Say that plainly.

DR. U: Genus is just the number of holes in a surface. A sphere has genus zero. A torus has genus one. As the hierarchy descends, the topology simplifies. The proton is the most complex. The electron is genus one. The neutrino is the simplest closed topology possible — genus zero.

DR. S: So the neutrino is spherical.

DR. U: In the topological sense, yes. Not a little billiard ball. A closed excitation without holes or knots. Topologically minimal.

Dr. S: And that is supposed to explain its weak coupling, neutrality, and lack of magnetic structure.

Dr. U: Exactly. The electron has charge and magnetic structure because it has a loop, and a loop supports persistent circulation. That is what the electromagnetic field couples to. A genus-zero structure has no loop and therefore no circulation of that kind. Without circulation, there is no charge and no magnetic structure. Without those, there is almost nothing for the electromagnetic sector to grip. The neutrino becomes a minimally burdened remnant that passes through ordinary matter almost untouched.

Dr. S: So the neutrino is not just light. It is topologically stripped down.

Dr. U: That is right. The endpoint of the descent sequence.

Now the shape of the hierarchy had become visible.
Trefoil. Torus. Sphere.
And because the sequence was visible, Dr. S could finally attack it properly.

Dr. S: All right. Let me state the attraction as clearly as I can. This is a cleaner particle story than the standard one in one important sense. It says the spectrum is not just a list of unrelated excitations with arbitrary masses and charges. It says stable structures simplify as the field regime descends.

Dr. U: Exactly.

Dr. S: That is attractive.

Dr. U: Yes.

Dr. S: But the danger is just as obvious. Topological stories are very easy to love before they are very hard to test.

Dr. U: Of course.

Dr. S: So where does yours become vulnerable?

Dr. U: In exactly the places where the geometry is supposed to constrain the observations. Mass ratios. Decay pathways. Weak-sector structure. And any signatures that depend on the particles' specific geometric forms rather than treating them as featureless points.

Dr. S: Good. Then let us talk about the uncomfortable case. The neutron.

The neutron belonged here because it interrupted the clean elegance of the sequence.
It was not the next lower topology.

It was a metastable composite.

That mattered.

DR. U: The neutron is not the next simple rung down. It is a strained composite in the same dense field regime as the proton.

DR. S: Figure-eight.

DR. U: Effectively, yes. A figure-eight-like composite organization — two lobes held together by the surrounding field environment. In the proton, you have one clean trefoil knot. In the neutron, you have something more like two opposed structures locked into one strained object.

DR. S: So it is not the ground state of heavy matter.

DR. U: No. The proton is. The neutron is metastable — supported under some conditions, unstable under others.

DR. S: Which is why free neutrons decay.

DR. U: Exactly. A free neutron outside an atomic nucleus lasts only about fifteen minutes on average. In standard physics, that is handled through the energetics of the weak interaction. In UFD, the point is geometric: the neutron's composite structure carries excess strain that the cleaner trefoil of the proton does not. Inside nuclei, the surrounding environment can help support that strain. In free space, that support is gone, and the structure eventually gives way.

DR. S: Into proton, electron, neutrino.

DR. U: Into trefoil, torus, and genus-zero remnant. Exactly.

That changed the mood of the discussion.

The sequence was no longer just a taxonomy.

It had become a decay logic.

DR. S: All right. That is stronger than just saying the particles have shapes. You are saying allowed decay channels reflect geometric simplification.

DR. U: Exactly.

DR. S: So the weak interaction begins to look less like a separate primitive force and more like topological failure in a metastable heavy structure.

DR. U: Yes.

DR. S: That is a dangerous claim.

DR. U: It has to be. Otherwise, the geometry is only decorative.

DR. S: Then walk me through the decay.

DR. U: The first step is dense-field relaxation. The composite figure-eight cannot sustain its configuration in free space, so it snaps through into the lower-strain proton configuration. One lobe survives as the trefoil ground state.

DR. S: And the rest?

DR. U: The expelled circulation cannot remain stably stored in the dense field layer. It descends into the lighter field, where it recoheres into the simplest stable structure that field can support. That is the electron.

DR. S: So the electron is not created from nothing.

DR. U: Exactly. It is recohered circulation shed by the collapsing heavy structure.

DR. S: And the neutrino?

DR. U: After the electron forms, there is still residual mismatch — some remaining strain that cannot stay stable even in the lighter field. So the system descends one level further, into the weakest-coupled layer, where the lowest-burden closed structure is the genus-zero neutrino. That is what carries away the final remnant.

DR. S: So the whole process is a cascade.

DR. U: Yes. A relaxation cascade across the hierarchy. The neutron cannot stay where it is. It has to shed excess burden, step by step, into lower field layers and simpler topologies.

For a moment neither of them spoke.
The discussion had now crossed a threshold.
This was no longer just a reinterpretation of one force.
It was a reinterpretation of one of the most familiar particle transformations in physics — beta decay — as a descent event across nested fields.

Dr. S broke the silence.

DR. S: Let me say what is attractive here. This does not merely rename the neutron decay products. It turns the channel itself into evidence of order. Proton, electron, neutrino is no longer just what happens. It becomes the direction in which the topology can simplify.

DR. U: Exactly.

DR. S: And it ties back to your earlier point about the proton. The trefoil is not chosen merely because it is knotted, but because it is the simplest stable topology whose threefold organization can answer to the proton's observed internal structure.

DR. U: Right. The hierarchy only matters if the geometry begins to answer to the observations.

DR. S: Which is actually a point in your favor. If the framework is going to be wrong, at least it will be wrong in a correlated way.

DR. U: That is one of the virtues of real unification.

That line hung in the air for a moment.
A weaker theory fails in separate compartments.
A stronger one fails along its joints.
Dr. S distrusted grand systems.
But he distrusted them less when they understood that principle.

So he gave the discussion its final test.

DR. S: Let me try the summary test.

DR. U: Go ahead.

DR. S: On your view, the particle spectrum is ordered by topological descent across nested fields. The proton is the low-strain trefoil ground state of heavy matter — chosen not merely because it is knotted, but because it is the simplest stable topology whose threefold organization can answer to the proton's observed internal structure when it is probed at high energies. The neutron is not a fundamental sibling but a metastable composite, a figure-eight-like configuration in the same dense field, stable inside nuclei but unstable in free space. The electron is a toroidal excitation in the lighter field — a doughnut-shaped vortex simple enough to support circulation, charge, and magnetic interaction without the knotted burden of heavy matter. And the neutrino is the topological endpoint of the descent, a closed genus-zero structure in the weakest-coupled layer, minimally burdened and barely coupling to anything else.

DR. U: That is fair.

DR. S: The attraction is obvious. The particle table stops looking like a zoo and starts looking like an ordered set of stable topologies.

DR. U: Yes.

DR. S: The danger is equally obvious. Once you say that, everything about masses, couplings, and decay paths has to begin lining up with the geometry.

DR. U: Exactly.

DR. S: Good. Then the burden is finally clear.

INTERLUDE

By the end of the exchange, the particle spectrum no longer looked flat. At least in the UFD picture, it had become a descent structure.

That mattered because topological language is cheap until it becomes selective. Once particle identity is tied to what different field layers can actually support, the geometry is no longer free to decorate the spectrum after the fact. It has to begin constraining which structures are stable, which are metastable, and which transformations are natural.

The neutron made that pressure unavoidable.

It interrupted the elegance of the sequence and, by doing so, made the sequence more interesting. It was not the next simple rung down. It was a composite under strain. Its decay therefore became more than a familiar reaction channel. It became the clearest early test of whether topological descent was only a pleasing picture or the beginning of a real physical ordering principle.

The attraction remained.

So did the cost.

And that was exactly where the argument needed to be.

TEMPORARY VERDICT

Dr. U succeeded in giving the particle hierarchy a real organizing principle. Topological descent made the spectrum look less like a list and more like a lawful sequence of stable structures, with the neutron functioning as the decisive complication rather than an embarrassment to be hidden.

Dr. S succeeded in sharpening the cost of that move. Once topology is made central, it cannot remain decorative. It has to constrain mass structure, decay logic, and weak-sector behavior in a disciplined way, and it has to do so without outrunning the present maturity of the framework's less-secured sectors.

So the conversation ends where it should.

IF PARTICLE IDENTITY REFLECTS WHAT EACH FIELD LAYER CAN STABLY SUPPORT, THEN THE NEXT QUESTION IS WHETHER THE WORLD'S MOST IMPORTANT CONSTANTS ALSO CARRY THAT SAME IMPRINT OF HIERARCHY.

CHAPTER 3

THE FINE-STRUCTURE CONSTANT, DENSITY SCALING, AND THE FIELD HIERARCHY

IS ALPHA A NUMBER, OR A CLUE?

The numbers that govern physics are among the strangest things in science.

Some are enormous in consequence but tiny in value. Planck's constant is so small that it only becomes visible at atomic scales. The gravitational constant is so weak that it takes the entire mass of the Earth to produce the pull we feel every day. Others are more mysterious still because they are pure numbers with no units attached to them at all. The fine-structure constant, which governs the strength of electromagnetism, is approximately 1/137. The proton is about 1836 times heavier than the electron. These are not just measured quantities. They are clues — or at least they look as though they ought to be.

Modern physics treats them as constants of nature: fixed numerical inputs that must be measured and inserted into the equations by hand. Once that is done, the results are extraordinary. The equations predict experiment with astonishing precision. But the deeper question remains. Why these values? Why 1/137 rather than 1/100 or 1/200? Why 1836 rather than 1000 or 3000? Why should the world contain just this numerical architecture rather than another?

The usual answer is pragmatic. The constants are what they are. Physics measures them, uses them, and moves on. From the standpoint of prediction, that posture works extremely well. From the standpoint of explanation, it leaves one of the deepest questions in the subject untouched.

Unified Field Dynamics begins from a more pointed suspicion. If reality is organized through a nested field hierarchy, then at least some of the constants should begin to reflect that hierarchy. They should not look like a flat list of unrelated gifts. They should look like traces left by a common architecture.

On that view, the fine-structure constant is not merely the abstract strength of electromagnetism. It is a clue to how energy descends from one field layer to the next. The speed of light is not a brute feature of spacetime, but the characteristic propagation speed of a real physical medium — the Universal Light Field. Planck's constant reflects the minimum quantum of circulation that the lighter field can sustain. The gravitational constant expresses the large-scale softness of a medium whose underlying stiffness is far greater than ordinary gravity suggests.

Not every constant is claimed to be fully derived, and some still require one-time normalization against experiment. But the ambition is definite: the constants should begin to look less like unrelated inputs and more like fingerprints of one field architecture.

By the third conversation, Dr. S had decided that ontology alone was no longer enough. A theory could sound deep for quite a while if it spoke in the right nouns — fields, coherence, geometry, topology, descent. But sooner or later every serious framework had to face the same question: where, exactly, does the architecture touch the numbers?

That was where he wanted to begin.

DR. S: All right. I understand the field hierarchy well enough now to see why it appeals to you. But we are past the stage where ontology by itself can carry the argument. Where does this architecture actually constrain the constants?

DR. U: Not all at once, and not all in the same way. Some constants are better treated as structural clues than others, and some relations are tighter than the absolute values themselves. The careful claim is not that everything is already derived. It is that several constants stop looking flat once the fields are treated as a hierarchy rather than as separate sectors.

DR. S: Good. That is already a better answer than triumphalism. So start with the obvious one: alpha. Why alpha?

This was the right place to begin.
A hierarchy that claimed to run deep into the structure of reality ought to leave traces in the pure numbers physics could not blame on human units.

DR. U: Because alpha is dimensionless, ubiquitous, and suspiciously central. A dimensionless number is a pure number with no units attached to it — no meters, no seconds, no kilograms. That matters because most physical constants change numerically when you change units, even though the physics stays the same. A dimensionless number does not. It is therefore much harder to dismiss as an artifact of convention. If the same dimensionless number keeps appearing at the boundaries between structure, propagation, and coupling, it begins to look less like a gift and more like a clue.

DR. S: A clue to what?

DR. U: To hierarchy. To the fact that not all fields carry the same physical burden, and that descent from one layer to another may come with a stable scaling cost.

DR. S: "Scaling cost" sounds promisingly dangerous. Explain it.

DR. U: In UFD, the fields are not just conceptually different. They differ in effective density, stiffness, and in the kinds of structures they can sustain. If a lighter communicative field emerges from a denser binding field, that descent should leave geometric ratios behind. Alpha is one candidate trace of that descent. Not yet a fully closed output in every detail, but no longer just an unexplained number either.

DR. S: So the claim is not that alpha has already been conjured from nothing.

DR. U: Correct. The claim is that alpha has stopped looking arbitrary.

That was worth lingering on.

A weaker alternative would have tried to impress by pretending total closure.

A stronger one distinguished between what the framework had already sharpened and what it still had to finish.

Dr. S pressed on.

DR. S: Fine. Then make it concrete. What, in your view, do the major constants become?

DR. U: They become physical expressions of the hierarchy rather than separately postulated numbers. The speed of light becomes the propagation speed of the Universal Light Field — how fast signals travel through the lighter layer. The gravitational constant becomes a macroscopic expression of the Universal Energetic Field's stiffness and pressure geometry. Planck's constant marks the minimum stable unit of circulation the lighter field can sustain. Alpha becomes a ratio tied to the cost of descent from the denser field to the lighter one. And mass ratios begin to reflect both topology and the price of crossing field tiers.

DR. S: That is an elegant inventory. But elegance is cheap.

Dr. U: Of course. The question is whether the reinterpretation constrains anything.

Dr. S: Does it?

Dr. U: Yes. For example, the hierarchy does not merely say that the fields differ. It says they differ in a specific scaling way. One of the clearer relations in the framework is the density-descent rule between the dense binding field and the lighter communicative field:

$$\frac{\rho_{\text{ULF}}}{\rho_{\text{UEF}}} \sim \alpha^3$$

where ρ_{UEF} is the effective energy density of the dense binding field and ρ_{ULF} is the effective energy density of the lighter communicative field.

Dr. S: And why the cube? Why α^3 rather than α or α^2?

Dr. U: Because the descent is three-dimensional.

Dr. S: Meaning?

Dr. U: The weakening is not just linear. It compounds across volume. If a structure scales down by one factor along one dimension, that is one thing. But if it opens into a lighter layer across all three spatial dimensions at once, the reduction scales as the cube of that factor.

Dr. S: Give me the picture.

Dr. U: Shrink the side of a cube to one-tenth, and the volume does not fall to one-tenth. It falls to one-thousandth. That is the logic here. The drop is volumetric, not merely proportional.

Dr. S: So the cube is doing real work.

Dr. U: Exactly. It is not decorative numerology. It is the claim that descent through the hierarchy is a three-dimensional density scaling.

Dr. S: Good. That is finally the kind of statement I was waiting for. Not just "the world is layered," but "the layering leaves ratios."

Dr. U: Exactly.

Dr. S: Fine. Then give me one place where that cubic descent is supposed to do real work. Not eventually. Now.

Dr. U: The clearest early example is the neutrino sector. If the electron belongs to the lighter communicative layer and the neutrino belongs to the still weaker coordination layer, then the same volumetric descent rule implies a cubic suppression of the neutrino mass scale relative to the electron.

DR. S: Say that more plainly.

DR. U: In the present framework, the neutrino mass is not left arbitrary. It scales as

$$m_\nu \approx m_e \alpha^3.$$

Starting from the electron mass, that places the neutrino in the sub-electronvolt range, at about **0.198** eV.

DR. S: So the cube is not just philosophical scenery. It is already being asked to constrain an actual particle scale.

DR. U: Exactly. And that is why the framework becomes vulnerable there. If the hierarchy is real, the cubic descent should show up not only in density language, but in the mass structure of the weakest-coupled excitations as well.

DR. S: And what does that buy you beyond a single number?

DR. U: It means the same ratio that organizes field descent should also organize the mass structure of particles that live in different field layers. Constants across domains stop behaving like strangers. If alpha appears in field descent, in inertial scaling, in the proton-electron mass ratio, and now in the neutrino scale, then the framework is not merely renaming known numbers. It is beginning to relate them through one underlying architecture.

That changed the feel of the conversation.

The issue was no longer whether UFD found the constants philosophically unsatisfying.

It had become a question about whether some of them could be reorganized by one underlying geometry.

Dr. S recognized the advance immediately, and also the danger.

DR. S: Let me say what worries me now. Once a theory starts treating constants as clues, it can become very seductive. Everything begins to look meaningful. Every ratio turns into evidence of hidden architecture.

DR. U: That is a real danger.

DR. S: So what keeps this from turning into numerology with fluid-dynamics vocabulary?

This was the right warning.

Alternative theories often become most vulnerable when they begin assigning meaning to famous numbers. Without strict limits, a search for structure quickly turns into numerology. Dr. S was demanding a standard severe enough to keep that from happening.

DR. U: Restriction. Cross-domain reuse. And failure conditions. A number counts as a clue only if it appears where the architecture says it should, in the form the architecture says it should, and with consequences that pressure other domains if it is wrong. Otherwise, it is just pattern hunger.

DR. S: Better. So alpha is not special because it is pretty. It is special because, in your framework, it recurs structurally.

DR. U: Right. And because it helps organize more than one relation without being introduced as a separate ad hoc constant.

DR. S: Including the proton–electron mass ratio?

DR. U: Yes. That is one of the places where alpha stops being merely suggestive and starts carrying physical weight.

The proton–electron mass ratio is one of the most striking numerical facts in ordinary matter.

Protons and electrons are the two most familiar stable charged constituents of atoms, yet the proton is about 1836 times heavier than the electron. Any theory that relates that number to deeper structure is making a serious claim.

Dr. S leaned forward slightly.

That was the point he had really been waiting for.

Not the philosophy of constants.

The arithmetic.

DR. S: Fine. Then let us go straight at it. If alpha is truly grounded, I do not want to hear only that it is "important" or "central." I want to know what work it actually does.

DR. U: Then start with the clearest case. In UFD, the proton–electron mass ratio is not treated as a brute numerical accident. It factorizes into two physically different pieces:

$$\frac{m_p}{m_e} \approx \frac{1}{\alpha} \, \xi_{\text{geo}} \approx 137 \times 13.4 \approx 1836$$

DR. S: Unpack that for me. What are those two factors actually doing?

DR. U: Different jobs. The $1/\alpha$ term is the cost of descent across field layers. The ξ_{geo} term is the extra cost of maintaining a more topologically demanding structure.

DR. S: More specifically.

DR. U: The electron is a simpler structure in a lighter field. The proton is a more heavily loaded structure in a denser one. So one part of the ratio reflects the burden of crossing layers at all. The other reflects the fact that a trefoil costs more to sustain than a torus.

That mattered because alpha was not being asked to explain everything by itself.
One factor measured the cost of descent across field layers.
The other measured the added burden of a more complex topology.
The ratio had been broken into two different physical jobs.

DR. S: So alpha is not doing the same work as ξ_{geo}.

DR. U: Exactly. And that is the point. If the whole mass ratio were explained only by knot geometry, alpha would still look imported. If it were explained only by field descent, topology would disappear. The framework says both burdens are real and separable. Alpha measures the descent cost. ξ_{geo} measures the topological cost. Together they give the full ratio.

That sharpened the conversation immediately.
Alpha was no longer only a clue in the abstract.
It had become part of a concrete structural identity.

Dr. S did not let that pass too easily.

DR. S: All right. Then tell me about ξ_{geo}, because right now that sounds like the place where the trick could be hiding.

DR. U: Fair. ξ_{geo} is not a fitted fudge factor. It is the topological impedance ratio — the difference in maintenance cost between the proton's trefoil-like structure and the electron's toroidal one.

DR. S: "Topological impedance" meaning what?

DR. U: The maintenance cost of localized topology. Not ordinary drag, not propagation loss, and not a simple volume ratio. A torus lets circulation flow smoothly around a loop. A trefoil does not. It forces the field through repeated crossings, tighter curvature, and more stored near-field stress. The torus is clean. The trefoil is choked. $\xi_{\mathbf{geo}}$ measures that extra burden.

DR. S: So the proton is not heavier just because it is "bigger" in some cartoon sense.

DR. U: No. It is heavier because it is topologically more expensive to sustain. In UFD, mass is not a measure of how much stuff is in a structure. It is a measure of how much coherence cost the field has to pay to keep that structure stable.

DR. S: And numerically?

DR. U: In the present treatment, $\xi_{\mathbf{geo}}$ is estimated at about 13.4. Combined with the field-loading factor $1/\alpha \approx 137$, that places the proton–electron mass ratio at the right scale:

$$\frac{m_p}{m_e} \approx \frac{1}{\alpha}\, \xi_{\mathbf{geo}} \approx 1836$$

This is not a completed first-principles closure of alpha. It is evidence that alpha is already embedded in a real architectural relation rather than floating as a decorative constant.

DR. S: Good. That is much stronger. Because now alpha is not merely "important." It is carrying one definite physical meaning.

DR. U: Yes. In the present framework, alpha is the dimensionless loading ratio that tells you how costly it is for coherent structure to step down into the lighter field sector. That is why it shows up both in the language of electromagnetic coupling and in the language of descent structure. The two descriptions are looking at the same physical process from different angles.

That was the conversation's real hinge.
Before that, alpha could still have sounded like a suggestive symbol.
Now it had been made to do work.
Not total work.
Not final work.
But real work.

Dr. S, being Dr. S, pushed on the exact place where the claim could still inflate.

DR. S: All right. But now the obvious caution. You still are not claiming the exact numerical value of alpha has been fully closed from first principles.

Dr. U: Correct. And that distinction matters.

Dr. S: Say it plainly.

Dr. U: The framework is strongest where it shows that alpha is structurally placed, cross-domain, and physically interpretable. It is weaker where full first-principles numerical closure is still pending. Alpha is not being treated as arbitrary. That is already stronger than standard practice. But it is not yet being presented as fully derived in every detail.

Dr. S: Better.

Dr. U: It has to be better. Otherwise, the theory weakens itself by overclaiming.

Dr. S: And the mass-ratio identity helps because it makes the incompleteness precise rather than vague.

Dr. U: Exactly. It shows what part of the result belongs to field descent, what part belongs to topology, and therefore what part still has to tighten if the framework is going to claim deeper closure later.

That answer improved the model more than any triumphant line could have done.
A weaker theory would have used alpha to sound profound.
A stronger theory used alpha to separate burdens.

Dr. S widened the frame one final time.

Dr. S: Let me step back. On your picture, the constants are not all on the same footing. Some remain closer to what physicists call normalization anchors — numbers that have to be fixed once by measurement so that everything else in the theory can be expressed relative to them. Some are already showing structural reuse — appearing in multiple places in ways that suggest they are tracking something real. And alpha is one of the places where the hierarchy begins to leave a trace.

Dr. U: Yes.

Dr. S: That is the gain.

Dr. U: Yes.

Dr. S: The danger is also clear. Once a theory starts reorganizing constants, it has to keep the line between clue and closure very clean.

Dr. U: Exactly.

Dr. S: And if that line blurs, it turns into numerology.

Dr. U: Right.

That was the center of the exchange.

Not whether UFD preferred a deeper explanation.

The issue was whether one of the most famous constants in physics was beginning to look like a structural trace of hierarchy rather than an unexplained gift.

That was a much harder claim.

And therefore a much more serious one.

Dr. S: Let me try the summary test. On your view, the constants of physics are not all on the same footing. Some remain normalization anchors — numbers fixed once by measurement. Some may eventually close more fully than they have so far. But several of the most important ones already look less like independent givens and more like traces of one field hierarchy. The speed of light becomes the propagation speed of the lighter field. The gravitational constant becomes a macroscopic expression of the denser field's stiffness. Planck's constant becomes a circulation or curvature-limit quantity. And alpha becomes a clue to field descent and electromagnetic loading that is connected to field hierarchy rather than brute numerical luck. More specifically, alpha is already doing one clear job: it appears as the field-loading factor in the proton–electron mass-ratio identity, while ξ_{geo} carries the topological impedance burden. So alpha is not yet fully closed, but it is no longer floating free.

Dr. U: That is fair.

Dr. S: That is a real gain.

Dr. U: Yes.

Dr. S: It is also a real exposure.

Dr. U: Yes.

Dr. S: Good. Then at least the issue is finally clear.

INTERLUDE

By the end of the third exchange, the conversation had moved from ontology to structure.

That mattered.

A theory can survive for a while on physical pictures alone. It cannot survive there indefinitely. Sooner or later, the architecture has to begin leaving numerical traces.

That was now the pressure on UFD.

Not that it must already have derived every constant from nothing. That would be an unreasonable demand, and in some cases a misleading one. The real pressure was narrower and more serious: do the constants begin to look related where standard practice treats them as separate? Do the ratios recur where the hierarchy says they should? Do the numbers begin to behave less like a list and more like evidence of common structure?

Alpha mattered because it moved the issue from general aspiration to a concrete burden. Once it entered as a field-loading factor in a quantitative mass-ratio relation, the theory had done more than say that the constant was philosophically interesting. It had started assigning it a physical job.

That did not finish the argument.

But it made the argument much harder to dismiss.

Temporary Verdict

Dr. U succeeded in moving the discussion of constants away from bare acceptance and toward structural interpretation. The key gain was not the claim that every number is already fully derived but the stronger claim that several major constants may be reorganized as traces of field hierarchy, scaling descent, and topology. Alpha became more than a suggestive symbol once it was given a definite architectural role as a field-loading factor, distinct from the topology-dependent burden carried by ξ_{geo}.

Dr. S succeeded in forcing the distinction that protects the model from overreach. A clue is not yet a closure. A geometric reinterpretation becomes scientifically serious only when it constrains multiple relations, survives cross-domain pressure, and remains vulnerable to failure.

So the conversation ends where it should.

IF THE CONSTANTS ARE BEGINNING TO LOOK LIKE FINGERPRINTS OF HIERARCHY RATHER THAN DISCONNECTED GIVENS, THEN THE NEXT QUESTION IS WHAT KIND OF DEEPER FIELD COULD IMPRINT THAT HIERARCHY IN THE FIRST PLACE.

CHAPTER 4

THE UNIVERSAL STRING FIELD AND THE COSMIC BLUEPRINT

THE SOURCE CODE OF PHYSICAL FORM

The question of why the universe has the form it does is one of the oldest questions in natural philosophy. It appears in Plato's search for deeper form behind the visible world, in Leibniz's principle of sufficient reason, and in modern physics whenever the laws and constants of nature begin to look less like brute facts and more like clues. Physics is remarkably good at telling us what the world does. It is less settled on why the world permits this kind of order at all.

Unified Field Dynamics asks that question directly. What if lawfulness is not primitive? What if the lower fields, constants, topologies, and coherence limits all inherit their structure from a deeper layer — one whose role is not to produce specific behavior itself, but to determine which kinds of behavior are physically possible in the first place?

In UFD, that deeper layer is the Universal String Field, or USF. The name is meant literally enough to matter. A guitar string can sustain some vibrational modes and not others. A drumhead can support some standing-wave patterns and forbid others. In each case, the structure of the instrument determines the set of stable forms that can exist within it. UFD proposes that reality itself has something like this structure at its deepest level. The USF is the harmonic substrate whose allowed modes determine which forms, topologies, and configurations can exist in the layers above it.

That is why the USF is not a rulebook imposed on reality from outside. It is the deeper physical medium whose internal geometry makes some patterns possible and others impossible. On this view, the laws of physics are not arbitrary commandments. They are the downstream consequence of what the harmonic structure of the USF permits.

From that substrate, the lower fields inherit their architecture. The Universal Energetic Field emerges as the dense binding layer. The Universal Light Field emerges as the lighter communicative one. The hierarchy below is therefore not just stacked mechanically. It is constrained from beneath by harmonic selection at the deepest level.

What made this chapter difficult was that it moved one level higher than most physics discussions usually go. The issue was no longer only what the world is made of, but why the world permits the kinds of stable form, law, and ratio that it does.

Maturity note. The USF is the most foundational and least empirically constrained layer of the field hierarchy. Its role is architectural: it provides the harmonic substrate from which the lower fields inherit their structure, topology, and coherence limits. Its effects are therefore indirect and downstream. For that reason, this chapter should be read as the philosophical completion of the field hierarchy rather than as a sector with independent quantitative benchmarks. Its burden is coherence with the rest of the framework, not the production of standalone experimental signals.

By this point, Dr. S no longer needed to be convinced that UFD was ambitious. What he wanted to know was whether it knew where its own ambitions had to stop. A theory could talk about deeper harmony, hidden order, and foundational structure for a very long time without becoming anything more than elevated rhetoric. If the USF was going to matter, it had to do more than sound profound. It had to explain something real.

That was where he began.

DR. S: Fine. We have pushed the burden steadily upstream. Constants are no longer supposed to be arbitrary. Topologies are no longer supposed to be accidental. Lower fields are no longer supposed to explain themselves. So now I want the obvious answer. What is the USF actually for?

DR. U: To explain why the lower fields have the structure they do.

DR. S: More specifically.

DR. U: Why stable forms are possible at all. Why the constants begin to look architectural rather than donated. Why topology is not just an inventory of lucky survivors. Why physical law is mathematical in a way that feels inherited rather than merely stipulated. And why the universe is ordered enough for the rest of the framework to make sense in the first place.

DR. S: So the USF is not there to add mystery. It is there to absorb mystery.

DR. U: Exactly.

DR. S: That is a good line.

DR. U: It is also the burden. If the USF does not actually reduce unexplained structure, then it becomes part of the unexplained structure.

That interested Dr. S. He mistrusted foundational language less when it came with a test like that: either it reduced mystery, or it became part of the mystery.

So he pressed on the place where that test would bite hardest. Law itself.

DR. S: All right. Then here is the obvious question. Why should I believe the universe has a blueprint at all? Maybe the laws are just brute facts. Maybe reality is simply one of the worlds in which these equations happen to hold. Physicists sometimes appeal to the idea that we live in one of many possible universes, and ours just happens to have the particular laws we observe. Why not accept that and stop looking for deeper reasons?

DR. U: It may be brute fact if one gives up early enough.

DR. S: That is glib.

DR. U: No. It is precise. "Brute fact" is often what we say when we stop asking why a lawful pattern is lawful. If constants, ratios, symmetry limits, and field relations keep looking harmonically structured — keep looking as if they belong together — then eventually the burden shifts. The burden is no longer only to ask why the world is lawful. It is to explain why we should stop asking.

DR. S: So for you, fine-tuning is not evidence of lucky parameters. It is evidence of hidden architecture.

Fine-tuning is the observation that many of the constants of nature appear to lie within narrow ranges compatible with stable atoms, stars, and life. Change the strengths of the basic interactions too far, and ordinary structure breaks down. The usual responses are familiar: brute fact, anthropic selection, or some larger multiverse in which many different combinations are realized. UFD treats the matter differently.

DR. U: Exactly. My claim is that all three of those responses accept arbitrariness too quickly. UFD proposes instead that the constants are architectural — that they look the way they do because the deepest field's harmonic structure permits some configurations as stable and excludes others. Fine-tuning, on this view, is not luck. It is a clue.

DR. S: Which means alpha, the speed of light, Planck's constant, the field ratios, curvature limits — all of that now belongs upstream.

DR. U: Yes. The earlier claims only become serious if the deepest layer is constraining them.

For a moment neither of them spoke. The issue had sharpened. This was no longer just about cosmology or field hierarchy. It was about whether the universe is lawful because it happens to be, or because deeper geometry makes other possibilities impossible.

Dr. S decided to make the challenge harder rather than softer.

DR. S: Fine. But "harmonic architecture" is still dangerously close to a slogan. Tell me what it means physically.

DR. U: Think of a vibrating plate.

DR. S: A Chladni plate.

A Chladni plate is a thin metal surface scattered with fine sand or powder. When the plate vibrates, the grains move until they settle into stable geometric patterns — rings, lines, starbursts, and combinations of them. The patterns are not drawn on the plate from outside. They emerge from the vibrational mode itself.

DR. U: Exactly. At first the motion looks messy. But once the mode settles, stable geometry appears. The pattern is not imposed from outside. It is selected by the vibration. UFD treats lower reality that way. The USF vibrates, and its stable harmonic modes define the attractors available to the lower fields. Form is not added later. It is inherited from what the deeper vibration allows.

DR. S: So the lower fields are not free to take arbitrary shapes.

DR. U: Right. That is the whole point. "Everything vibrates" is not a theory. The question is what the vibrations constrain. My claim is that the USF's vibrations constrain geometry, topology, and which patterns of coherence are physically admissible. The lower fields inherit those limits. They do not invent them.

That helped. Not because it proved anything. Because it changed the character of the claim. The USF was no longer a hidden metaphysical layer whispering order into reality. It had become a physical answer to a physical question: why are some forms stable, some ratios persistent, and some structures lawful across scales?

Dr. S saw the gain immediately. And the risk.

DR. S: All right. Let me say what is attractive here. The Chladni picture makes the framework more intelligible. Form is not a mathematical ghost. It is stable geometry selected by vibration. Fine. But analogies are cheap too. How do you get from this deeper harmonic layer to the world we actually see?

DR. U: Through projection and shedding.

DR. S: Meaning?

DR. U: The Universal Energetic Field — the denser layer that carries heavy matter and nuclear binding — is projected from the USF. The deeper harmonic layer sets the admissible geometric templates, and a founding cosmological event localizes that order into a bounded substantive domain. I will say more about that cosmological event later. For now, the point is simpler: the deeper field constrains the forms, and the lower substantive field realizes them.

DR. S: And from there?

DR. U: Then the Universal Light Field — the lighter layer that carries electromagnetic waves and quantum coherence — is shed from the UEF. The lighter layer separates from the denser one because the denser layer cannot hold all its energy in one configuration. The result is a nested architecture: the USF provides the harmonic templates, the UEF provides the substantive energy that realizes them, and the ULF provides the communicative layer through which the world's parts interact.

DR. S: Form, substance, communication.

DR. U: Yes. In compressed form: form first, then substance, then communication.

DR. S: So this is not a bottom-up pile of unrelated sectors.

DR. U: No. It is a cascade. The deeper field does not merely sit beneath the others. It constrains what they are allowed to become.

Dr. S tapped the table lightly. He was not persuaded, but he could now see the shape of the claim more clearly. That made it easier to test.

So he asked the next hard question.

DR. S: Let me put the standard objection cleanly. Physics has heard this promise before. String theory also said reality was fundamentally vibrational. It also promised a deeper source of law and unification. Why is this not just a more intuitive rerun of the same instinct?

String theory is the most famous modern attempt to ground particle physics in a deeper vibrational picture. Its mathematical richness is undeniable. But it has also been criticized for relying on extra dimensions, for allowing an enormous number of possible vacuum states, and for not uniquely fixing the low-energy physics we actually observe. That is why the comparison matters.

DR. U: Because the instinct is similar, but the ontology is not.

DR. S: More specifically.

DR. U: String theory kept the vibrational intuition but paid for it with ontological permissiveness: extra dimensions, huge landscape freedom, and weak low-energy uniqueness. UFD keeps the vibrational intuition but rejects that permissiveness. The world stays four-dimensional — three of space and one of time, the same dimensions we inhabit. The USF is not a machine for generating endless compactifications. It is a physically real harmonic constraint layer whose stable modes tighten what lower reality can be.

DR. S: So your complaint is not that string theory was too ambitious.

DR. U: No. It is that it became too permissive. A good foundational theory should constrain, not multiply. If the USF becomes a landscape generator rather than a constraint generator, it fails in exactly the same way.

That mattered. Dr. U was strongest whenever he made UFD more vulnerable, not less. A weaker theory would have wanted to inherit string theory's aura without inheriting its burden. A stronger one made the comparison exact and then sharpened the difference where it counted: admissibility rather than permissiveness.

Dr. S went next where he always went when a deep layer was invoked. Fine-tuning.

DR. S: All right. Then let us talk about the constants directly. Standard physics does not really explain why they are what they are. Sometimes people appeal to anthropic selection. Sometimes to a multiverse landscape. Sometimes they shrug and treat the constants as brute facts. What does your picture say?

DR. U: That the constants are not arbitrary knobs. At least not in the deepest sense. They are reflections of admissible geometry across the nested fields. The USF's harmonic structure determines which ratios, which couplings, which field relationships can sustain themselves and which cannot. The constants we observe are the ones that can actually exist stably in a universe where the USF's harmonic order is what it is.

DR. S: So alpha, the speed of light, Planck's constant, the field ratios—

DR. U: All become pieces of one harmonic architecture rather than independent dials.

DR. S: Are you claiming all of them are fully derived?

DR. U: No. That would be careless. The stronger claim is that they stop looking like disconnected donations from nowhere and start looking like inherited constraints inside one structured world. The earlier alpha discussion is an example. Alpha is not yet fully derived from first principles, but it is no longer floating free either. It is doing structural work in ways that would be difficult to explain if it were just a brute input.

DR. S: Which is less triumphant and more serious.

DR. U: It has to be more serious. If all the USF does is provide prettier language for the same arbitrary inputs, then it has not done its job.

That answer interested Dr. S. He distrusted grand language less when it came with restraint of that kind. It suggested the theory knew the difference between architecture and completion.

So he turned to another place where deeper field theories usually break down: the vacuum.

Modern quantum theory does not treat the vacuum as mere emptiness. Even "empty" space carries residual activity — zero-point structure — and that activity shows up in real effects such as the Casimir effect. But once one treats the vacuum as energetic, a major problem appears. If all vacuum activity gravitated in the same way as localized energy, the resulting gravitational effect should be fantastically larger than what is actually observed. That mismatch is one version of the cosmological constant problem.

DR. S: Fine. Then what becomes of the vacuum? Because every deep field theory eventually has to answer that.

DR. U: The vacuum hum remains real.

DR. S: But?

DR. U: It does not gravitate in the naive way. The USF's background vibration is not the same kind of thing as a local pressure gradient in the UEF. They live at different levels of the hierarchy and do different physical jobs.

DR. S: So zero-point activity is reclassified as non-gravitating background order.

DR. U: Exactly.

DR. S: Which means the cosmological constant problem gets reframed rather than solved numerically.

DR. U: Better: dissolved at the right level. The mistake is to assume that uniform background harmonic activity must gravitate the way a local concentration of energy gravitates. A violin string vibrating in a standing mode is carrying real vibrational energy, but that is not the same thing as a localized pressure excess in the air around it. In UFD, gravity belongs to pressure geometry in the lower substantive field — to localized energy concentrations in the UEF — not to the mere fact that the deepest field hums.

DR. S: So vacuum activity remains real, but the inference to catastrophic gravity does not survive the ontology shift.

DR. U: Exactly. You cannot calculate the gravitational effect of vacuum energy by pretending the vacuum is just a very thin soup of localized energy. It is something else. It is a background mode of a deeper field, and background modes do not gravitate the way localized concentrations do.

That was a strong moment. Not because it settled the cosmological constant problem, but because it showed the kind of move the USF was supposed to make. It did not merely add a deeper layer. It changed which downstream inferences were even allowed.

Dr. S saw that. And then he made the conversation harder again.

DR. S: All right. Suppose I grant that the USF helps explain lawfulness, fine-tuning, and maybe even why uniform vacuum activity should not gravitate. There is still a deeper problem. Standard entropy language says the universe runs downhill. Disorder increases. Thermalization spreads. Eventually, in the bleakest picture, the universe approaches heat death. Why should I think your harmonic picture does anything but rename that fact?

DR. U: Because the usual entropy picture is better at tracking local mess than global order.

DR. S: Go on.

DR. U: Thermalization is real. Entropy increase is real. But that is not yet the whole story. The deeper question is whether the field architecture underneath those local processes is settling into more stable patterns or less stable ones. In UFD, time does not only move toward breakdown. It also moves toward more stable coherence basins — regions of configuration space where lawful organization is self-sustaining.

DR. S: So you are not denying thermal disorder.

DR. U: No. I am putting it in context. What looks like disorder at one level can still belong to a deeper movement toward harmonic settlement at another.

DR. S: That sounds suspiciously convenient.

DR. U: Only if it stays vague. Think of the Chladni plate again. The grains look chaotic while the mode is settling. A microscopic observer might see nothing but local scrambling. But the system as a whole is moving toward a stable pattern selected by the underlying vibration. Local scrambling and global selection are not the same thing.

DR. S: So the arrow of time becomes a movement toward deeper order rather than simple decay.

DR. U: Toward deeper coherence, yes. Not every local system becomes more ordered. Entropy genuinely increases in many places. But the universe as a whole does not have to be read as a long slide into formlessness. Its deeper tendency can still be toward harmonic settlement.

DR. S: And this is where the Geometric Coherence Force enters.

DR. U: Exactly. The GCF is not an extra force added alongside the others. It is the general tendency of reality to settle into more stable harmonic configurations. That is why lawful structure can look directive without being externally imposed. The laws are not commandments handed down from outside. They are the stable patterns that survive because they are the ones the harmonic structure allows.

That changed the tone of the exchange. The USF was no longer only explaining why form exists. It was beginning to explain why time itself need not be read simply as the erosion of all form. A weaker theory would have turned that into poetry. A stronger one kept it attached to mechanism: vibration, stable attractors, coherence basins, and the distinction between local disorder and global settlement.

Dr. S did not let it drift any farther than that.

DR. S: Let me be fair. There is something genuinely attractive here. A great deal of foundational physics ends with resignation: the laws are what they are, the constants are what they are, the vacuum is what it is, and perhaps one day a deeper formalism will connect the pieces. You are offering something stronger. You are saying the pieces belong together because deeper harmonic structure makes them belong together.

DR. U: Yes.

DR. S: And the danger is obvious.

DR. U: Of course.

DR. S: The danger is that "deep harmonic structure" becomes a permission slip for every unanswered question. Every puzzle becomes evidence for the USF. Every surprise becomes proof of hidden architecture. That is exactly how foundational theories turn into unfalsifiable metaphysics.

DR. U: That is exactly the danger.

DR. S: So what keeps this from happening?

DR. U: The same thing that keeps any foundational layer honest. It has to reduce independent assumptions, constrain lower structure, and expose itself to failure through the things it claims to ground. If the lower fields, constants, topologies, cosmology, and coherence limits do not become tighter under the USF — if the framework's architecture does not actually reduce the number of things we have to take as brute inputs — then the

USF is not the source code of reality. It is just one more invisible sector. The test is not whether the USF sounds profound. The test is whether the rest of physics starts making more sense once you assume it.

That was the center of the discussion.
Not whether a blueprint sounded elegant.
Whether a blueprint reduced arbitrariness.

Dr. S still mistrusted deep foundations. But he mistrusted them less when they came with a criterion sharp enough to kill them.

So he gave the discussion its final test.

Dr. S: Let me try the summary test. On your view, the USF is not a decorative final layer. It is the deepest physically real harmonic field in the hierarchy, and its stable modes define the admissible forms of lower reality. The UEF is projected from it, the ULF is shed from that substantive domain, and the world we observe is the outcome of form, substance, and communication ceasing to be separate. The Chladni-style picture is not there for poetry but to say that stable form is selected by vibration rather than imposed from outside. Fine-tuning becomes evidence of inherited architecture rather than lucky parameters. Uniform vacuum activity becomes non-gravitating background order rather than an automatic cosmological catastrophe. And the arrow of time itself begins to look less like simple loss and more like movement toward deeper coherence under the Geometric Coherence Force.

Dr. U: That is fair.

Dr. S: The attraction is obvious.

Dr. U: Yes.

Dr. S: The danger is equally obvious.

Dr. U: Yes.

Dr. S: Good. Then the burden is finally clear.

INTERLUDE

By the end of the exchange, the question was no longer whether the universe might have a deeper layer.

It had become whether that deeper layer actually made the world less arbitrary.

That mattered.

A hidden field earns nothing merely by being deep. It earns its place only if
it reduces the number of things that must still be accepted as unexplained:
lawful constants, stable topologies, inherited coherence, admissible form,
and the strange fit between mathematics and physical reality. In the UFD
picture, the USF is where those mysteries are supposed to meet. If they do
not meet there, then the framework has only pushed the unexplained one
step farther back.

The Chladni analogy mattered because it made the deepest claim physically
legible. Form was no longer a ghostly abstraction hovering over matter. It
was the stable pattern selected by vibration. That did not prove the USF
was real. But it made the claim much clearer: not that reality is mysteriously
aesthetic, but that lawful structure may be the downstream expression of a
real vibratory order.

The gain was obvious. So was the burden.

TEMPORARY VERDICT

Dr. U succeeded in making the USF intelligible as a real explanatory layer
rather than a decorative depth claim. Lawfulness, fine-tuning, inherited
form, the nested cascade of lower fields, and the GCF as a physical basis
for natural law all became more coherent once the deepest layer was treated
as a real harmonic constraint field rather than as a mathematical
afterthought.

Dr. S succeeded in forcing the line that protects the whole discussion from
overreach. A blueprint theory does not earn its place by sounding
profound. It earns it only if it reduces arbitrariness, constrains lower
structure, and survives incompleteness without pretending that every
downstream law has already been fully closed from first principles.

**IF LAWFUL FORM IS INHERITED AND THE LOWER FIELDS ARE
PROJECTED FROM A DEEPER HARMONIC ORDER, THEN THE NEXT
QUESTION IS NO LONGER WHY THE UNIVERSE IS INTELLIGIBLE. IT IS
WHAT THE FIRST SUBSTANTIVE FIELD ACTUALLY DOES.**

THAT IS WHERE GRAVITY BEGINS.

CHAPTER 5

GRAVITY, INERTIA, AND DARK MATTER

CURVATURE, IMPEDANCE, AND CURRENT

Gravity is the oldest problem in physics and, in a strange way, still one of the least understood. We can predict with astonishing precision how an apple falls, how a planet orbits the Sun, and how a spacecraft moves between worlds. But the deeper question — what gravity actually is — remains open. Physics has achieved extraordinary predictive power without yet arriving at a completely satisfying physical picture.

The modern story begins with Isaac Newton. In 1687, Newton proposed that every piece of matter in the universe attracts every other with a force proportional to their masses and inversely proportional to the square of the distance between them. The law worked. It explained planetary motion, tides, projectiles, and much of classical mechanics. But Newton himself was uneasy with the idea that bodies could act on one another across empty space without any intervening medium. His law described how gravity behaved, but not what gravity was.

In 1915, Albert Einstein proposed a radically different answer. Gravity, he said, was not a force in the ordinary sense but the curvature of spacetime itself. Matter and energy bend the geometry around them, and other bodies move along the natural paths that curved geometry makes available. The result was one of the most beautiful and successful theories in science. General relativity explained the precession of Mercury, the bending of starlight, black holes, gravitational waves, and much more.

And yet some of the deepest questions remained.

One is inertia: the resistance of a body to acceleration. Why is a truck harder to push than a bicycle? Why should the same quantity that measures a body's resistance to being pushed also measure its response to gravity? Einstein showed that inertial and gravitational mass are equivalent, and that equivalence became one of the great structural facts of modern physics. But the physical reason for it remained obscure.

Another is dark matter. Beginning in the twentieth century, astronomers found that galaxies and clusters moved as though there were far more

62

gravitating matter present than telescopes could see. Fritz Zwicky saw it in galaxy clusters. Vera Rubin saw it in spiral galaxies. Outer stars orbited too quickly. Clusters held together too strongly. Gravitational lensing revealed more mass than visible matter could account for. The standard framework answered by adding an unseen ingredient: dark matter, a form of matter inferred through gravity but not yet identified directly.

The resulting picture is formidable. It works across enormous scales, from solar-system mechanics to cosmology. But its ontology remains divided. Gravity is geometry. Inertia is accepted as basic. Dark matter is an unseen supplement. Three phenomena that seem as though they might belong together are described in three different ways.

Unified Field Dynamics proposes a different physical picture. Start not with empty space, but with a real medium. In UFD, the universe is not a void with geometry imposed on it from outside. It is a structured field environment — specifically, the Universal Energetic Field, the denser layer that carries heavy matter and pressure geometry. A coherent structure within that medium — a star, a planet, a galaxy, any system held together as one organized whole — changes the pressure pattern of the surrounding field. Nearby bodies respond to that pattern.

The key idea is a pressure gradient. A pressure gradient is simply the rate at which pressure changes over distance. High pressure on one side, lower pressure on the other, and anything caught in the medium is driven from one toward the other. Winds, currents, and fluid flow all depend on this principle. The force is not acting from nowhere. It is the medium itself carrying bodies along the gradient it creates.

UFD proposes that gravity is exactly this kind of process. A massive structure creates an organized pressure pattern in the surrounding field. Other structures do not feel a mysterious pull across emptiness. They respond to the pressure gradient of the real medium around them and accelerate along it. Gravity, on this view, is not space bending in a void. It is motion produced by pressure geometry in a real field. The mathematics of relativity can still remain valid as an effective description. What changes is the ontology. Curvature is no longer fundamental. It becomes the large-scale appearance of a deeper pressure geometry.

Inertia becomes the same physics seen from another side. To accelerate an object is to force a coherent structure to change its motion through a coupled field environment. That reconfiguration costs work. Inertia, in UFD, is not a brute property of matter. It is the resistance of organized structure to being accelerated through the coupled medium. That is why inertial and gravitational mass are equivalent: they are two expressions of the same deeper coupling.

Dark matter is the third face of the same mechanism. If gravity is pressure geometry in a real medium, then the pressure pattern surrounding visible matter need not fall off in exactly the Newtonian way at large scales. At solar-system scales the departures may be negligible. But at galaxy scales the organized pressure structure can extend farther than visible matter alone would suggest. In that case, what standard cosmology interprets as a dark halo need not be a separate population of invisible particles. It may instead be the extended pressure geometry induced by visible coherent systems in the surrounding medium.

That is the unifying ambition of the chapter. Gravity, inertia, and dark matter are no longer treated as separate problems. They become three aspects of one ontology: pressure-gradient flow in the dense field, impedance in the lighter field, and large-scale organized gravitational behavior without unseen particulate halos.

The gain is obvious. So is the burden. If one medium is doing this much work, it must do it with discipline. The theory must match what is observed, not merely offer a more tangible story.

What made this conversation especially important was that it gathered three problems normally treated in different languages into one proposed mechanism. If UFD can do that without losing empirical discipline, it gains a great deal. If it cannot, the whole picture becomes too expensive to keep.

Maturity note. The pressure-geometry interpretation is not merely conceptual. At galaxy scale, the framework now has a quantitative anchor. Using the SPARC database of disk-galaxy rotation curves, UFD's finite-radius cored pressure model reproduces the observed rotation-curve phenomenology and outperforms the canonical NFW halo in much of the sample. A companion weak-lensing analysis suggests that the same SPARC-calibrated velocity scale reaches consistently into the galaxy-scale lensing regime without separate retuning. That is a genuinely strong result for an alternative framework.

But it is not yet full closure. Disturbed systems, galaxy mergers, and cluster-scale structure — including the kinds of systems standard dark-matter arguments treat as

---

TECHNICAL NOTE — PRESSURE GEOMETRY AND EFFECTIVE CURVATURE

The simplest picture is this: gravity is not space bending in a void, but motion caused by pressure differences in a real medium. Massive structure changes the pressure pattern of the surrounding field, and nearby bodies respond to that flow.

More precisely, UFD relocates the ontology of relativity. Curvature is treated as an effective macroscopic description of organized pressure geometry rather than as the fundamental substrate itself. The mathematics of general relativity can still describe observed gravitational behavior correctly where it has been tested. What changes is the physical interpretation of what those mathematics are describing.

At galaxy scale, the framework's pressure picture is no longer only qualitative. The finite-radius cored profile reproduces the observed rotation-curve phenomenology of disk galaxies, and the same pressure-geometry picture reaches outward into a first galaxy-scale weak-lensing consistency test without parameter readjustment. That is stronger than reinterpretation alone. But the lensing sector remains only partially developed, and the constitutive treatment of disturbed systems, mergers, and clusters still has to be earned explicitly. UFD is not claiming to have solved gravity and dark matter in full. It is claiming that the pressure-geometry picture has earned a place among serious alternatives.

By the fifth conversation, Dr. S had become less interested in whether UFD was physically imaginable.

He wanted to know whether it could bear weight.

Gravity was the obvious place to ask.

Not because it was easy.

Because it was not.

DR. S: Fine. Let us go directly at the hardest thing. In standard relativity, gravity is curvature. Whatever its conceptual price, that framework has earned enormous trust. So what, exactly, are you replacing it with?

DR. U: A pressure flow in a real medium.

DR. S: Say that more carefully.

DR. U: In UFD, matter is a coherent vortex embedded in the UEF. That vortex imposes boundary constraints on the surrounding medium and creates a localized pressure deficit. Gravity is the resulting inward acceleration of the field — and of structures coupled to that flow. Not a primitive pull. Not spacetime curvature as the ontological starting point. A pressure-gradient response.

DR. S: So a whirlpool picture.

DR. U: In the relevant sense, yes. The analogy is not decorative. It is causal.

DR. S Then let me state the obvious objection. If the medium is as dense and fundamental as you say, why does everything not collapse catastrophically?

DR. U: Because density is not the same thing as force. Uniform pressure does not generate motion. Pressure gradients do. That distinction is basic hydrodynamics, and it matters here. A uniform UEF does not gravitate. Only the gradients induced by coherent excluded geometry do.

DR. S Good. That is at least a real answer.

This was not a minor clarification.
It was the difference between a catastrophic medium theory and a viable one.
If sheer density produced gravity automatically, the whole picture would collapse under its own weight.
UFD survived this first objection only by making gradients, not mere fullness, do the causal work.

That answer did more than block an objection.
It revealed the kind of ontology UFD was willing to risk.
The medium was no longer bookkeeping.
It was the actor.

That made the picture more vivid.
It also made it easier to kill if it failed.

Dr. S noticed that too.

DR. S All right. Suppose I grant, for the sake of argument, that gravity is a real pressure current. Then the next question is immediate. What is inertia?

DR. U: Resistance from the other side of the same structure.

DR. S: Meaning?

DR. U: Gravity and inertia arise from the same coherent vortex, but they couple to different field layers. Gravity is how the vortex shapes the dense field. Inertia is how the lighter propagation field resists changes in the vortex's motion.

DR. S: So gravitational mass and inertial mass are not two unrelated kinds of mass.

DR. U: No. They are two ways one structure engages two nested media.

DR. S: Go on.

DR. U: When a vortex moves uniformly through the ULF, the surrounding organization can remain symmetric. But when it accelerates, it has to build and carry a distortion halo with it. That costs work. Inertia is the cost of reconfiguring the field around accelerated motion. It is not friction. It is not dissipative drag. It is a conservative impedance effect.

DR. S: Hydrodynamic added mass.

DR. U: Exactly.

DR. S: So the mystery of inertia becomes a field effect.

DR. U: Yes. A real one.

"Added mass" is a standard idea from fluid dynamics. When an object accelerates through a fluid, it does not only move itself; it must also accelerate some of the surrounding medium. Dr. U was borrowing that intuition here, but applying it to a field rather than to an ordinary liquid.

For the first time in the exchange, Dr. S looked less skeptical than interested.
Not convinced.
Interested.

That mattered more.

DR. S: Let me see if I understand the symmetry you think you have found. The same coherent object produces gravity because it shapes the UEF around it, and inertia because it disturbs the ULF when accelerated.

DR. U: Exactly.

DR. S: Which means the equivalence principle is not an axiom.

DR. U: Not on this view. It is a structural consequence.

DR. S: Explain that without poetry.

DR. U: If the same topology determines both the gravitational source geometry in the UEF and the impedance geometry in the ULF, then gravitational and inertial mass scale together for a given class of matter. Not because nature simply declares them equal, but because the same vortex is doing both jobs.

DR. S: So a boulder and a pebble fall alike because the stronger pressure response of the larger structure is matched by the stronger inertial burden of that same structure.

DR. U: Exactly. Current and resistance scale together.

This mattered because the equivalence of gravitational and inertial mass is one of the deepest structural facts in physics. In standard theory, it is built into the framework at a very basic level, but without explanation. UFD was trying to make it emerge from one shared geometry instead.

That was the first real deepening of the conversation.
Gravity alone might still have been dismissed as a fluid metaphor.
Gravity and inertia, tied to one geometry across two fields, made the framework much more ambitious.

And ambition meant burden.

Dr. S pressed there next.

DR. S: Fine. Then what happens to G?

This question mattered because G, the gravitational constant, is one of the basic numerical anchors of physics. If UFD could reinterpret it as a trace of medium stiffness rather than a primitive gift, the framework would gain real explanatory depth. If it could not, the rhetoric of pressure geometry would remain incomplete.

DR. U: It stops being fundamental.

DR. S: Of course it does.

DR. U: It has to. If gravity is a pressure-gradient effect in a real medium, then G should reflect the medium's stiffness and characteristic scale, not appear as an unexplained gift. In UFD, G is the large-scale manifestation of UEF stiffness. Its smallness does not mean gravity is metaphysically weak.

It means the medium is extraordinarily stiff, and ordinary gravitational effects are the macroscopic residue of that much deeper scale.

DR. S: So the weakness of gravity becomes a scale issue, not an ontological mystery.

DR. U: Exactly.

DR. S: And you actually have a relation in mind.

DR. U: Yes. The framework links the UEF surface-tension constant, k, and a characteristic microscopic coherence scale. That relation is not there for atmosphere. If it fails once the scale is fixed, this sector of the framework fails with it.

DR. S: Good. That is the kind of vulnerability a real alternative needs.

The conversation had now become expensive in exactly the right way.
A medium story is easy to love when it remains qualitative.
It becomes much harder — and much more serious — when it links constants, scales, and mechanisms in a way that can go wrong.

Dr. S then turned to the obvious wider burden.

Dark matter.

Dark matter is one of the hardest places to replace modern physics. The standard view did not invent it as a decorative idea. It was forced on physicists by observations: galaxies rotate too fast for the visible matter alone to hold them together, clusters bind more tightly than their visible mass should allow, and gravitational lensing reveals more mass than any telescope can see. Any alternative framework has to do more than offer a different picture. It has to recover at least the strongest parts of that observational burden.

DR. S: All right. Let us go to the astronomical problem. The standard view says galaxies and larger systems behave as though there is far more gravitating matter than we can see. You are not allowed to wave that away just because you dislike halos.

DR. U: I agree.

DR. S: Good. Then what replaces them?

DR. U: Larger-scale pressure geometry in the same medium.

DR. S: Meaning?

DR. U: Visible baryonic structure is not sitting inside an otherwise neutral background. Its coherent organization shapes the surrounding UEF over large scales. The resulting pressure-gradient structure can alter orbital behavior without requiring a separate population of collisionless halo particles.

DR. S: So dark matter becomes an inference error.

DR. U: Not exactly. The observations are real. The ontology behind the inference changes. Standard theory sees unexplained gravity and adds unseen matter. UFD sees the same behavior and asks whether the medium's extended pressure structure is already enough.

DR. S: And you think that can flatten rotation curves?

DR. U: In the relevant regime, yes. That is one of the framework's explicit burdens.

DR. S: Burden, not victory.

DR. U: That is how it began. But it no longer stands only as a conceptual burden. At galaxy scale, the pressure picture can now be formulated as a finite, cored profile and tested against real rotation-curve data.

DR. S: Tested is one thing. Has it actually done anything?

DR. U: Yes. In the SPARC analysis, the cored pressure profile reproduces the rotation-curve phenomenology across the sample and outperforms the canonical NFW halo under matched statistical treatment.

SPARC is a major galaxy-rotation dataset often used to test models of galactic gravity. NFW refers to the standard cuspy halo profile used in dark-matter modeling. The importance of Dr. U's claim was not only that UFD sounded cleaner, but that it had entered the same comparison space as the canonical halo picture.

DR. S: That is a stronger claim than "the ontology is cleaner."

DR. U: It has to be. Otherwise, the ontology earns nothing.

DR. S: Fine. But rotation curves are where many alternatives sound strongest. Does the picture reach beyond them at all?

DR. U: At a first level, yes. A companion weak-lensing comparison suggests that the same SPARC-calibrated velocity scale extends into the galaxy-scale lensing regime without being retuned there.

DR. S: Suggests, not settles.

DR. U: Exactly. That distinction matters. It is a first galaxy-scale consistency test, not yet a full constitutive lensing theory.

DR. S: So you are no longer only saying that the ontology is cleaner. You are saying it has begun to touch the phenomenology directly, and not only in-sample.

DR. U: Exactly. The point is no longer merely that extended medium structure sounds more physical than invisible halos. The point is that the same medium structure can be asked to recover the observed orbital behavior of real galaxies and then be carried outward, at least in a first way, into galaxy-scale lensing.

DR. S: That is stronger.

DR. U: It should be.

DR. S: But let me say the obvious next sentence for you. Galaxy-scale rotation curves and a first lensing consistency test are still not the whole dark-matter burden.

DR. U: Of course not.

DR. S: Lensing in full detail. Disturbed systems. Mergers. Cluster-scale structure. Systems where the standard view will say the case for collisionless matter becomes harder to avoid.

DR. U: Yes. Those remain major pressure points. The galaxy-scale result matters because it shows the medium picture can do more than redescribe a problem. The weak-lensing result matters because it shows the picture is no longer confined to one chosen battlefield. But the larger burden does not disappear with those gains.

DR. S: Good. Because otherwise this would just be one more theory that wins where it chooses the battlefield.

DR. U: Agreed.

DR. S: Let me sharpen the status, then. Rotation curves are now a quantitative anchor. Galaxy-scale weak lensing is a first consistency test. Disturbed systems, mergers, and clusters remain open.

DR. U: That is the careful version.

DR. S: Good. Then at least the maturity line is no longer vague.

DR. U: It cannot be vague.

Weak lensing matters because it tests gravity through light, not only through orbital motion. A theory can sometimes fit rotation curves where it chooses the battlefield. Extending the same structure into lensing is therefore a more serious step.

Dr. S did not smile, but something in his posture shifted.
He was not less skeptical.
He was more focused.

That was progress.

DR. S: Here is what I think the real issue is. Your picture is clearly more physically imaginable. Gravity becomes current instead of curvature. Inertia becomes impedance instead of primitive resistance. Dark matter becomes medium structure instead of invisible particles. That is all attractive.

DR. U: Yes.

DR. S: But attraction is cheap. The standard framework bought its place by carrying a huge empirical burden. So the question is not whether your ontology feels better. The question is whether one medium can really bear all the weight you are putting on it.

DR. U: That is exactly the question.

DR. S: Good.

DR. U: And I would add one thing. If one medium can bear that weight, then much of what currently looks like evidence for extra hidden substance may really be evidence that our ontology has become too permissive.

DR. S: That is a serious accusation.

DR. U: It has to be. Otherwise, there is no reason to make the replacement.

That was the deepest point of the exchange.
Dr. U was not merely offering a more tangible picture.
He was claiming that some of modern gravity theory's most successful supplements might be signs of ontological overgrowth rather than discovery.

Dr. S was not merely defending orthodoxy.
He was insisting that a more satisfying mechanism earns nothing unless it survives the burdens that made the standard one formidable.

That was exactly where the argument needed to be.

DR. S: Let me try the summary test.

DR. U: Go ahead.

DR. S: On your view, gravity is a pressure-gradient response of the UEF induced by coherent topological structure, not primitive spacetime curvature. Matter is a vortex that creates localized pressure deficits in a real medium. Uniform background density does not gravitate; only gradients do. Inertia is the hydrodynamic impedance of the ULF to acceleration, not an intrinsically mysterious property of matter. The equivalence between inertial and gravitational mass emerges because the same coherent structure couples to two different fields in two different ways. And dark matter is not a halo of undiscovered particles but the larger-scale pressure geometry of the medium surrounding visible systems — a claim that now has a quantitative galaxy-scale anchor in rotation curves, a first consistency test in galaxy-scale weak lensing, and still carries major unresolved burdens at the level of disturbed systems, mergers, and cluster structure.

DR. U: That is fair.

DR. S: The promise is clear. So is the cost of failure.

DR. U: Yes.

DR. S: Good. Then we know what has to survive.

INTERLUDE

By the end of the fifth exchange, gravity, inertia, and dark matter no longer looked like separate problems.
At least in the UFD picture, they had become three expressions of a single claim: coherent structure in nested real fields produces pressure gradients in the dense field, impedance in the lighter field, and large-scale, organized gravitational behavior without requiring unseen particulate halos.

That is a strong unifying move.
It is also a dangerous one.
Because the question is no longer whether gravity can be made more tangible. It is whether one medium can truly carry explanatory weight that standard physics presently distributes across curvature, inertial law, and dark matter.

The gain is now sharper than before.

The medium picture no longer lives only at the level of ontology. It now has a quantitative anchor at galaxy scale, and it has begun to reach, in a first consistency-test way, beyond rotation curves into galaxy-scale weak lensing. That does not yet make it a closed alternative. But it does mean the dark-matter reassignment is no longer merely conceptual.

The cost is just as sharp. Galaxy-scale success is not full closure. A replacement picture still has to survive disturbed-system lensing, mergers, cluster-scale superposition, and the broader gravitational burden without quietly reintroducing what it set out to eliminate.

The gain is obvious.

So is the cost.

And that was exactly where the conversation needed to be.

Temporary Verdict

Dr. U succeeded in making the merged picture physically coherent and empirically exposed. Gravity became pressure-gradient current, inertia became field impedance, and dark matter became extended medium structure rather than missing particles. Gravity became pressure-gradient current, inertia became field impedance, and dark matter became extended medium structure rather than missing particles.

Dr. S succeeded in holding the decisive line of pressure. Those gains matter, but they do not yet settle the case. A medium-based replacement earns its place only if it preserves the tested weak-field successes, links constants and scales without hidden freedom, and survives the harder burdens still ahead — disturbed systems, mergers, clusters, and full constitutive lensing — without quietly reintroducing what it set out to eliminate. So the conversation ends where it should.

IF GRAVITY AND DARK MATTER ARE REALLY EXPRESSIONS OF ONE STRUCTURED MEDIUM, THEN THE NEXT QUESTION IS WHAT THAT SAME MEDIUM HAS BEEN DOING TO LIGHT ALL ALONG. BEFORE ONE CAN TELL A DIFFERENT COSMIC HISTORY, ONE HAS TO KNOW WHETHER REDSHIFT ITSELF STILL MEANS WHAT MODERN COSMOLOGY SAYS IT MEANS.

CHAPTER 6

REDSHIFT, TIME DILATION, DARK ENERGY, AND THE HUBBLE TENSION

DID THE UNIVERSE EXPAND, OR DID THE MEDIUM EVOLVE?

Light carries information about where it has been. That is one of the deepest facts about how astronomy works. When we look at a distant star or galaxy, we are not only seeing a point of light. We are seeing a record: of the object's chemical composition, its temperature, its motion, and the medium the light has crossed on its way to us. That record is encoded in wavelength. Every chemical element leaves its own spectral fingerprint, and any systematic change in those wavelengths tells us something about the history of the light's journey.

The most famous such change is redshift. Light from distant galaxies reaches us with its wavelengths stretched toward the red end of the spectrum. Every spectral feature is displaced to a longer wavelength than it had when it was emitted. The usual analogy is the Doppler effect in sound. An approaching ambulance sounds higher because its sound waves are compressed; a receding one sounds lower because they are stretched. Light behaves similarly. Light from something moving toward us is blueshifted. Light from something moving away is redshifted.

In the 1920s, Edwin Hubble found that the light from most distant galaxies was redshifted, and that the amount of redshift tended to increase with distance. Nearby galaxies showed smaller shifts. More distant galaxies showed larger ones. That relation became known as Hubble's law, and it became one of the foundations of modern cosmology.

The standard interpretation is specific. Cosmology does not usually treat those galaxies as simply flying through a fixed space. It treats the redshift as the result of metric expansion: the stretching of space itself. On that reading, galaxies are not moving apart the way fragments from an explosion move through air. The space between them is growing, and the galaxies are carried along by that growth. Once redshift is read that way, a great deal follows. Run the process backward, and the universe appears to converge toward a much hotter, denser past. That became the basis of the Big Bang picture.

Other pieces of the modern cosmological story are built on the same interpretation. Type Ia supernovae appear to fade more slowly at high redshift, by the same factor that cosmic expansion predicts. The Hubble constant becomes the rate at which the cosmic scale factor changes over time. And dark energy is introduced to explain why the redshift-distance relation implies not just expansion, but late-time acceleration.

All of this gathers into a single narrative: an expanding universe with a measurable history of growth, change, and acceleration. The narrative is mathematically powerful and empirically rich. But it depends on one decisive reading of the evidence: that cosmological redshift is primarily the fingerprint of expanding space. If that reading is correct, modern cosmology has earned a remarkable degree of unity. If it is wrong at its core, the consequences do not stop with one datum. They spread outward through the whole story.

Unified Field Dynamics proposes a different reading of what redshift actually is. In UFD, light does not travel through an empty space whose geometry is stretching. It travels through a real medium — the Universal Light Field, the lighter communicative layer introduced in earlier chapters — and that medium has its own history. As light propagates across cosmological distances, it couples to the evolving medium it passes through. One consequence of that transport is wavelength stretching. From the observer's end, that stretching can look exactly like the redshift standard cosmology attributes to expanding space.

This is meant as a physical claim, not a verbal re-labeling. In UFD, redshift is the cumulative effect of propagation through an evolving Universal Light Field. The transport is not free. Light interacts with the medium it moves through, and that interaction has observable consequences. One is wavelength stretching. Another is temporal stretching: the same transport process that shifts wavelengths can also stretch the timing of arriving signals. On that view, supernova time dilation does not require a separate cosmological cause. It is part of the same propagation history.

The reinterpretation extends farther. What standard cosmology reads as evidence of accelerating expansion — the dimness of distant supernovae relative to a simpler expansion law — UFD reads as the observational imprint of continued ULF maturation. The medium itself is not static. Its transport properties evolve over cosmic time, and those changes alter the relationship between distance, wavelength, and observed brightness. Dark energy, in that reading, is not a new cosmic substance. It is a sign that the medium through which light has been traveling has a history of its own.

The Hubble tension then takes on a different meaning as well. In the standard picture, different methods of measuring the Hubble constant are supposed to converge on one true expansion rate. But they do not. Early-universe inferences based on the cosmic microwave background yield a lower value, while late-universe distance-ladder measurements yield a higher one. In standard cosmology, that discrepancy is a puzzle. In UFD, it is a clue. Different methods are not necessarily failing to recover the same number. They may be sampling different transport regimes — different histories of interaction between light and the evolving medium.

That is why this chapter matters. The issue is no longer only what gravity is. It is what light has been doing all along.

Maturity Note. *In the technical corpus, this chapter is best read as a constrained transport alternative rather than as a fully closed replacement for the standard cosmological pipeline. UFD has already developed the pressure-geometry interpretation of gravity to the point where specific quantitative tests can be made against galaxy-scale rotation curves and first weak-lensing comparisons. The cosmological sector remains earlier in maturity. It identifies a physical mechanism that could, in principle, account for redshift, time dilation, dark energy, and the Hubble tension without requiring metric expansion, but it has not yet produced a complete quantitative replacement for the full Lambda-CDM framework. The chapter should therefore be read as an ontological alternative with testable implications, not as a finished cosmological closure.*

The link to the previous discussion is not accidental. If gravity is a real current in a real medium — pressure geometry in the Universal Energetic Field — then light cannot be treated as though it travels through a world with no material history of its own. What pressure geometry does to matter locally, medium evolution may do to light across vast distances. The two pictures are not independent. They are two expressions of the same larger claim: that the universe is not empty space with phenomena laid into it, but a structured medium whose history helps shape everything that happens within it.

That was why the burden now shifted.
The question was no longer what gravity is.
It was what light has been doing ever since.

Dr. S had no intention of letting the argument proceed on atmosphere alone.

DR. S: Fine. We have now made the medium real. Good. Then the next question becomes unavoidable. If the same field architecture gives you gravity, what does it do to light over cosmic distances?

DR. U: In UFD, it stretches it.

DR. S: Not because galaxies are receding through space.

DR. U: Not fundamentally, no.

DR. S: Then let us be clear about the stakes. In modern cosmology, redshift is not just one observation among others. It is the optical backbone of the expansion story. If you reinterpret that, you are not merely revising one mechanism. You are putting pressure on the whole history built on it.

DR. U: Exactly. That is why the mechanism has to be stated carefully.

DR. S: Then state it carefully.

DR. U: In UFD, redshift is cumulative wavelength stretching in the Universal Light Field. The relevant quantity is not recessional velocity in the standard cosmological sense, nor a changing metric scale factor. It is a local stretching kernel associated with ULF transport and background geometry. As light propagates, that effect accumulates. So the observed redshift is the integrated result of real transport through a real medium.

DR. S: So not an old-fashioned tired-light story.

DR. U: Correct. Not random scattering. Not a simple fading mechanism. Not a hand-waving claim that light "just loses energy." A constitutive transport law in a real field.

That mattered.

A great many bad alternatives to expansion had died by sounding physical without becoming precise. Dr. U's advantage here was not that UFD made redshift feel more tangible. It was that it gave the stretching a definite physical carrier.

Dr. S pressed on.

DR. S: All right. Then what is the actual law supposed to be?

DR. U: Schematically, the idea is that redshift accumulates exponentially along the propagation path. If you integrate the local stretching contribution from source to observer, you get a closed-form result:

$$z = e^{Hd} - 1$$

where H is the local stretching rate set by ULF transport and background geometry.

DR. S: Exponential. Not linear with a correction. Let me make sure I follow what you are claiming, because this is bigger than it sounds. In the standard picture, redshift grows roughly in proportion to distance for nearby objects. The relationship only starts to bend at very large distances, and when it bends, it bends because the expansion history bends. You are saying the exponential is not an effect of cosmic expansion. It is built into the travel law itself.

DR. U: Yes. The exponential is there from the start. Standard cosmology reaches something like that only through the assumptions of metric expansion. UFD gets there directly, from the propagation law.

DR. S: And the Hubble radius?

DR. U: It changes meaning. In standard cosmology, it is tied to expansion. In UFD, it becomes the characteristic distance over which light undergoes one significant exponential step of stretching in the medium. Same scale. Different physics.

DR. S: So one Hubble distance means one natural step in the transport medium.

DR. U: Exactly. A material scale, not a spacetime boundary.

That was the first real gain of the conversation.
Not proof.
A cleaner picture of the observable.

The Hubble scale is one of the most quoted numbers in cosmology. It is used to talk about the age of the universe, the expansion rate, and the size of the observable domain. To say that the same number can be re-read as a property of the propagation medium rather than as a feature of expansion is not a small move. It does not change the measurement. It changes what the measurement is of.

The Hubble scale had stopped being a kinematic bookkeeping device and started behaving like a property of the medium itself. That alone did not make UFD right. But it made the disagreement sharper.

Dr. S moved immediately to the standard objection.

Supernova time dilation.

DR. S: Fine. A lot of bad cosmologies say light simply changes as it travels. They usually die on supernova time dilation. What happens here?

DR. U: In UFD, the time dilation is not an extra burden added after the redshift. It is the same process.

DR. S: Explain.

DR. U: If the wave packet's wavelength is stretched by the evolving ULF, then its temporal structure is stretched with it. The signal arrives both redshifted and broadened because the same field process governs both. Spectral stretch and temporal stretch are not two separate cosmological mechanisms. They are one transport effect seen two ways.

DR. S: So on your view, the supernova time-dilation law is built into the same medium story that produces redshift.

DR. U: Exactly.

DR. S: That at least avoids the most embarrassing tired-light failure mode.

DR. U: It has to.

Dr. S nodded once.
Not convinced.
But the target had become cleaner, and that was already progress.

The tired-light reference was not casual. Earlier alternatives to cosmic expansion often tried to explain redshift by saying that photons simply lost energy along the way. Those theories failed badly when supernova light curves were measured in detail. Distant supernovae do not merely arrive redshifted. Their temporal profiles arrive stretched by the same factor. Standard expansion predicts that naturally. Most tired-light models could not recover both effects from one mechanism. That is what made Dr. S's nod meaningful. Any transport-based reinterpretation of redshift enters the same dangerous neighborhood immediately. The first question is always whether it can explain supernova time dilation without bolting on a second story afterward.

He pushed next on the deeper physical link.

DR. S: All right. You said a moment ago that the same vortex action that gives you gravity also gives you this propagation effect. I want that said more plainly.

DR. U: Good. In UFD, coherent structures do two things. First, they organize pressure geometry in the UEF. That gives you gravity. Second, they shed phase into the communicative field. That changes the large-scale state of the ULF. Over cosmic distances, light propagating through that changing medium accumulates stretch.

DR. S: So local vortex organization gives one effect in the substantive field and another in the communicative field.

DR. U: Exactly. Gravity is the pressure-current side of the story. Redshift is the phase-stretching side. Same architecture. Different domain.

DR. S: That is one of your stronger moves.

DR. U: It should be. Otherwise, gravity and cosmological propagation would still be two unrelated mysteries.

That changed the feel of the conversation.

Redshift was no longer a detached cosmological trick.
It had become the long-range optical consequence of the same medium dynamics already used to explain gravity.

That was a real gain in unity.

Dr. S then moved to the next standard ingredient.

Dark energy.

DR. S: Fine. Then tell me what dark energy is.

DR. U: A misreading of medium evolution.

DR. S: Meaning?

DR. U: In UFD, recession is not the primary event in the first place. Light traveling through the cosmos is being stretched by the medium it passes through, and that medium is not the same at every stage of cosmic history. The propagation field matures. Later photons travel through a more

developed ULF than earlier ones did. So light from more distant sources
accumulates more stretching — not because anything is accelerating
outward, but because the medium itself has changed along the way.

Dr. S: So acceleration is not galaxies flying apart ever faster. It is the
observational effect of light traveling through a medium whose properties
have evolved.

Dr. U: Exactly.

Dr. S: Which means dark energy is not a new substance at all.

Dr. U: Not in UFD. It is the name we gave to a gap in our reading of the
data. Once the data are read differently, the gap closes, and there is no
substance left to find.

That was a strong claim.
And because it was strong, it needed discipline.
Dr. S supplied it immediately.

Dr. S: Let me state the mainstream case as fairly as I can. Dark energy was
not invented because cosmologists enjoy multiplying mysteries. It was
forced by a redshift-distance relation that curved the wrong way for a
matter-only universe inside the standard expansion framework. Standard
cosmology responded by adding a cosmological constant or something
close to it. Ugly, perhaps, but disciplined.

Dr. U: Or disciplined within a specific ontology.

Dr. S: Fair correction. But that is the point. If you remove expansion and
dark energy together, you are claiming that a huge amount of apparent
necessity was really interpretive entanglement.

Dr. U: Yes.

Dr. S: That is a dangerous claim.

Dr. U: Good.

Dr. S: You keep saying that as though danger were a virtue.

Dr. U: In this context, it is. A replacement cosmology that does not put
much at risk is not a replacement cosmology.

Dr. S did not like that answer.
Which was one reason it worked.

A weaker theory would have tried to sound careful by risking little. A stronger one accepted that replacing several standard ingredients with one medium-based mechanism increased both the gain and the burden.

That led naturally to the next pressure point.

The Hubble constant itself.

In standard cosmology, the Hubble constant is the number that tells you how fast the universe is expanding. It is usually written as a velocity per unit distance. The current measured value sits somewhere around seventy.

DR. S: All right. Then what is the Hubble constant on your view?

DR. U: Not a universal expansion rate. An effective stretch rate of the ULF.

DR. S: So not the time derivative of a cosmic scale factor.

DR. U: No. More like the coherence-growth or elastic-relaxation rate of the propagation medium itself.

DR. S: Which means you do not expect every observational method to return one single universal value.

DR. U: Correct.

DR. S: So now we come to the Hubble tension.

DR. U: Exactly.

The Hubble tension is one of the most active controversies in modern cosmology. Local distance-ladder measurements give a value around seventy-three. Early-universe inferences based on the cosmic microwave background give a value closer to sixty-seven. The gap is small in absolute terms but large compared with the quoted uncertainties, and years of refinement have not removed it.

In the standard framework, this is a serious problem because the Hubble constant is supposed to be one universal property of one expanding universe. Dr. U's position is different. If the Hubble parameter is really a property of transport, not expansion, then different methods need not be converging on one number at all. They may be probing different transport regimes.

DR. S: So your claim is that what standard cosmology treats as a crisis is really a signature.

DR. U: Yes — though carefully stated. UFD does not say the tension by itself proves the standard model wrong. It says the tension is exactly the sort of thing one would expect if the underlying question had been misframed — if there were no single universal expansion parameter in the first place.

DR. S: Meaning different measurements are probing different transport regimes.

DR. U: Exactly. Local ladder measurements are more influenced by local cavity-weighted or boundary-weighted transport. Intermediate probes sample a broader intergalactic regime. And CMB-based inferences fold global transport together with assumptions inherited from the standard cosmological mapping. On this view, the mismatch is not an embarrassment to one universal number. It is evidence that there was never only one number to recover.

That was one of the model's stronger moves.
A weaker theory would have used the Hubble tension as a slogan.
A stronger theory used it as a structural promise that still had to earn its precision.

Dr. S made sure that point did not get lost.

DR. S: Let me be very careful here. You are not saying you have already delivered a fully closed quantitative solution to the Hubble tension.

DR. U: No.

DR. S: Good.

DR. U: The careful version is that UFD naturally anticipates an early/late inference divergence because different pipelines sample different transport environments. But the precise numerical separations remain targets of modeling rather than fully closed derivations.

DR. S: Better.

DR. U: It has to be said that way. Otherwise, the argument overstates itself.

DR. S: And what would failure look like?

DR. U: Failure would look like this: the ULF stretching law fails to match the observed curvature of the Hubble diagram, fails to reproduce the supernova time-dilation relation, violates spectral or surface-brightness constraints, or fails to support a coherent layered account of the Hubble split. In that case, the medium-transport picture fails on its own terms.

That answer improved the model.
A theory becomes more believable when it knows exactly how it could lose.

Dr. S then widened the frame one final time.

DR. S: Let us step back. Standard cosmology did not earn its place only because of redshift. It earned its place because once redshift was read as expansion, a whole history came with it: time dilation, relic background, large-scale growth, late-time acceleration, and the rest. If you replace that with medium evolution, do you realize how much structure you are taking on?

DR. U: Completely.

DR. S: Then tell me why it is worth taking on.

DR. U: Because the standard story pays for its success with a growing inventory of abstractions: metric expansion, dark energy, inflation, dark matter, singular origin. UFD asks whether one medium-based history can do more of that work with fewer primitive categories.

DR. S: Which is either economy or overreach.

DR. U: Exactly.

DR. S: And the only way to tell is to see whether the medium history tightens under pressure.

DR. U: Yes.

That was the center of the exchange.
Not whether the alternative sounded more intuitive.
Whether it could bear enough observational weight to justify replacing several standard ingredients with one evolving medium.

Dr. S still trusted the standard picture more than the replacement.
Dr. U still thought the standard picture had mistaken a successful map for a final ontology.

Nothing about that had changed.
But the disagreement was now much cleaner.

DR. S: Let me try the summary test.

DR. U: Go ahead.

DR. S: On your view, redshift is cumulative wavelength stretching in an evolving ULF, not recession in expanding space. The low-redshift Hubble law survives only as the small-distance limit of a more general constitutive transport law. Time dilation is not a separate cosmological effect but the temporal side of the same stretching process. Dark energy is the observational imprint of continued ULF maturation rather than a repulsive cosmic substance. And the Hubble tension is not a failure of one expansion history to return one number, but evidence that different observational methods are probing different transport regimes.

DR. U: That is fair.

DR. S: The attraction is obvious.

DR. U: Yes.

DR. S: The danger is equally obvious.

DR. U: Yes.

DR. S: Good. Then the deeper stakes are finally clear. If the standard interpretation of redshift is wrong at the root, modern cosmology cannot simply be patched at the edges. Its whole narrative structure has to be reconsidered — not just expansion and dark energy, but the meaning of the background radiation and the kind of beginning the universe is supposed to have had.

DR. U: Exactly. Once propagation is reinterpreted, cosmic history is reinterpreted with it.

INTERLUDE

By the end of the exchange, the issue was no longer whether redshift could be given a more tangible picture.

It had become whether one medium-based transport mechanism could really replace several of the standard model's most important cosmological functions at once.

That mattered.

In the standard picture, redshift, time dilation, apparent acceleration, and the Hubble constant all belong to one expansion history. In the UFD picture, they are reorganized as consequences of one evolving propagation medium. That is a genuine gain in unity if it works. It is also a genuine increase in exposure.

The argument became strongest where it stayed disciplined. UFD does not need to claim that every detail of the Hubble split is already quantitatively closed. It only needs to claim something harder and more interesting: that a layered medium should naturally generate different effective Hubble-like inferences, and that this can be tested as a transport picture rather than defended as a philosophical preference.

And now the next consequence was unavoidable.

If redshift no longer tells the story of expanding space, then the rest of cosmology cannot remain untouched. The question of origins changes with it. So does the meaning of the cosmic microwave background.

The gain remained obvious.
So did the burden.
And that was exactly where the discussion needed to be.

TEMPORARY VERDICT

Dr. U succeeded in giving the cosmological propagation story a genuine physical core. Redshift became constitutive ULF stretching, time dilation became part of the same mechanism, dark energy became medium maturation, and the Hubble tension became layered transport rather than a cracked expansion history.

Dr. S succeeded in holding the line that matters. A replacement cosmology does not earn its place by sounding more tangible. It earns it only if one medium-based architecture can carry the observational burden now distributed across expansion, dark energy, and Hubble-scale inference—and do so without overstating the present level of quantitative closure.

IF LIGHT CAN CHANGE ITS WAVELENGTH BECAUSE THE MEDIUM ITSELF EVOLVES, THEN THE NEXT QUESTION IS NO LONGER WHETHER A DIFFERENT COSMOLOGICAL STORY IS POSSIBLE.

IT IS WHAT KIND OF BEGINNING THAT NEW STORY PERMITS.

CHAPTER 7

GENESIS EVENTS, SUPERMASSIVE BLACK HOLES, AND ISLAND GALAXIES

A UNIVERSAL EXPLOSION, OR A BOUNDED ACT OF CREATION?

Every culture has asked where the universe came from. For most of human history, the answers were mythological or theological. Modern cosmology is unusual because it offers a scientific origin story — and not just an origin story, but one that is supposed to explain almost everything we later see through one continuous unfolding history.

The standard account is the Big Bang cosmology. About 13.8 billion years ago, the observable universe was in an extremely hot, dense state. From that state, it expanded and cooled. As the universe cooled, particles formed, then nuclei, then atoms. Light that had once been trapped in the early plasma was finally released, producing the cosmic microwave background. Over immense spans of time, small density variations grew into stars, galaxies, and large-scale cosmic structure. In this picture, the universe we see today is the late result of one universal early event.

The strength of that story is not only that it offers a beginning. It is that it ties many later observations into one arc: the cosmic microwave background, the abundance of light elements, galaxy formation, cosmic redshift, and the ages of stars all become parts of one continuous history.

Unified Field Dynamics proposes a different kind of beginning.

In UFD, the universe does not begin as one universal explosion that produces everything at once. It begins with what the framework calls a Genesis Event: a sudden organizing transition in the deeper field hierarchy that produces a bounded region of structured energy. In the fuller picture, Genesis has two levels. First comes a founding cosmological phase, in which the harmonic order of the deeper field precipitates into a substantive energetic domain. Then comes a local phase, in which vortex seeds within that larger environment organize matter into bounded island systems — what we observe as galaxies.

That difference matters because it changes the basic cosmological unit. In the standard picture, galaxies are late products inside one universal history.

In UFD, galaxies are not merely late products. They are primary bounded worlds with their own founding events, their own boundaries, and their own central organizing structures.

That is where supermassive black holes enter the story.

In standard astronomy, the supermassive black hole at the center of a galaxy is usually treated as the result of gravitational collapse and later growth. But their formation remains one of the harder open problems in cosmology, especially when very massive black holes appear surprisingly early in cosmic history. UFD proposes a different reading. The central black hole is not the late corpse of a collapse process. It is the original organizing seed of the galactic region itself. On this view, the galaxy grows around the central engine rather than the black hole appearing only after the galaxy is already in place.

Every galaxy, then, becomes a bounded cosmic archipelago: a local world with its own Genesis history, its own central seed, its own boundary, and its own maturation. What standard cosmology calls "the universe" becomes, in UFD's reading, one such bounded history among many rather than the whole of reality. The larger picture is not one cosmic beginning followed by everything else. It is a deeper ordered medium within which many local births can occur.

This is the chapter's most ambitious reinterpretation. It is not merely a new reading of one observation. It is a proposal about what kind of universe there is in the first place. Is there one universal history unfolding from a single beginning, or many bounded local histories distributed through a deeper cosmic architecture?

That was why this conversation raised the cosmological burden to its highest level so far. The issue was no longer how one sector of physics should be reinterpreted, but what kind of universe physics is actually describing.

Maturity Note. The Genesis Event, the island-galaxy model, and the larger cosmological architecture are structured proposals with identifiable physical mechanisms. But they are not yet observationally closed in the way the standard cosmological pipeline is closed. UFD has specified how bounded local genesis would work in principle, how central black holes would function as organizing seeds, and how a larger ordered environment could distribute many local histories. It has not yet produced a full quantitative cosmological model that can be matched observation by observation against Lambda-CDM. This chapter should therefore be read as a constrained alternative cosmological picture rather than as a finished replacement. Its burden is coherence with the rest of the framework, plausibility in the face of existing observations, and identification of the tests that would distinguish it from the standard account.

Cosmology is where a replacement theory stops being local. One can argue about gravity and still leave the larger sky mostly intact. One can reinterpret quantum theory and leave cosmic history untouched. But once the disagreement becomes one of origin and structure, the burden changes. A new ontology no longer has the luxury of solving one puzzle at a time. It has to say what kind of universe there is in the first place.

Dr. S had been waiting for that pressure from the beginning.

DR. S: Fine. We have delayed this long enough. Up to now, you have been replacing sectors one at a time — gravity here, particles there, propagation elsewhere. But cosmology is less forgiving than that. What, exactly, happened at the beginning?

This was the decisive question.
Many alternative theories revise one mechanism while leaving the larger cosmological story intact.
UFD was doing something riskier.
It was proposing a different kind of beginning altogether.

DR. U: In UFD, the careful answer is that the beginning was not a universal explosion into expanding space. It was a local coherence event.

DR. S: Local?

DR. U: Yes. A Genesis Event. A bounded quench in the deepest field layer that precipitates a structured energetic domain.

DR. S: Already I can hear the standard objection. The Big Bang is not just a picturesque explosion story. It is a highly constrained early-universe framework tied to background radiation, expansion, abundances, and structure formation. If you replace it, you inherit all of that burden.

DR. U: Completely.

DR. S: Good. Then start at the root. What is a Genesis Event?

DR. U: A sudden organizing event in the deepest field that creates a bounded region of dense energy. Standard cosmology starts with a hot dense state that expands. UFD starts with a coherence transition that produces a structured region — one with its own boundary, its own internal pressure geometry, and its own history. The universe is not an explosion. It is a founding event.

DR. S: So not one universe born everywhere at once, but a bounded cosmic birth.

DR. U: Exactly.

DR. S: And what actually comes into being?

DR. U: The UEF first. That is the substantial medium of mass, gravity, and large-scale pressure geometry.

DR. S: Emerging from the USF.

DR. U: Yes. The USF supplies the inherited harmonic structure. The UEF is the first substantive domain that can bear pressure, structure, and later matter formation.

DR. S: So the field hierarchy is doing cosmological work now, not just particle work.

DR. U: It has to. Otherwise, the hierarchy is ornamental.

That immediately made the claim clearer.
The issue was not merely whether one origin story sounded more intuitive than another.
It was whether the universe begins as one everywhere-at-once event or as many bounded births inside a larger ordered medium.

Dr. S pushed where the alternative story became most exposed.

Structure.

DR. S: Fine. Then what kind of structure does this event create? Because "bounded energetic domain" is still too abstract.

DR. U: A seeded island system.

DR. S: Meaning?

DR. U: The Genesis Event is not only the appearance of a medium. It is the appearance of a domain with internal geometry and external limit. The system develops around a central vortex seed. In the fuller story, what standard cosmology would treat as a later supermassive black hole is not an accidental late product. It is one of the earliest organizing structures in the island's maturation.

DR. S: So the galaxy begins centrally, not diffusely.

DR. U: In the relevant sense, yes. A galaxy is not just matter collecting inside a preexisting metric background. It is a bounded coherence domain organizing around a central seed within a structured medium.

DR. S: And the supermassive black hole is there from the start?

DR. U: Not necessarily in its final mature form from the first instant, but yes in the architectural sense: the central fracture or vortex seed is early and load-bearing. In UFD, supermassive black holes are long-term coherence engines, not merely late accidents of collapse.

This mattered because early supermassive black holes are one of the standard cosmological pressure points. In the conventional picture, they can appear surprisingly mature surprisingly early. UFD was not treating that as an anomaly to be patched later. It was trying to build black holes into the original architecture of galaxy formation itself.

A weaker theory would have left black holes as secondary objects because they were awkward to integrate.
A stronger one made them central.

DR. S: Then tell me what a black hole is on your view.

DR. U: Not what the textbooks usually describe. In standard physics, a black hole is treated as the end state of collapse — a region where matter has fallen inward so completely that the equations break down at a point of infinite density, hidden behind an event horizon. Even physicists working inside that framework will often say the singularity itself is not really physical. It is a sign that the theory has stopped working.

DR. S: And in UFD?

DR. U: A black hole is a real physical structure, not a breakdown. It is a region where pressure and loading become so extreme that ordinary outward channels close off and the system reorganizes around a deep central vortex. There is no point of infinite density. There is a stable, highly structured core, surrounded by a boundary where the field can no longer support ordinary outward flow.

DR. S: So not an infinitely dense point.

DR. U: No. The whole purpose of the framework is to replace breakdown with structure. In standard physics, when the equations stop making sense, you say a singularity is there. In UFD, you say the model is incomplete and ask what real structure is actually present. A black hole is not the place where intelligibility ends. It is one of the most organized structures in the universe. It is a coherence engine.

DR. S: And that matters cosmologically because—

DR. U: Because if the central object is an early organizing engine rather than a late accident, then galaxy formation changes meaning. In the standard picture, the puzzle is how very large black holes appear quickly enough inside young galaxies. In UFD, the central seed belongs to the architecture from the beginning. The island matures around it.

DR. S: So the chronology is almost reversed.

DR. U: In the important sense, yes. The black hole is not merely something a galaxy eventually produces. It is one of the structures through which the galaxy becomes itself.

DR. S: Which would make surprisingly early supermassive black holes less surprising.

DR. U: Exactly. And also very large black holes in comparatively small hosts, or systems where the central engine seems overdeveloped relative to the surrounding galaxy. In the standard story, those can look premature. In UFD, they are natural expressions of local Genesis under different loading conditions.

DR. S: And the jets?

DR. U: Easier to read physically as well. If the central object is a coherence engine, then jets are not just an awkward byproduct. They are coherence exhaust: organized release from an overdriven central structure.

DR. S: That at least gives them a place in the architecture rather than treating them as decoration.

DR. U: Yes. And it helps explain why the black hole and the host galaxy look so deeply linked. In UFD they are not two later-joined objects. They are parts of one island history.

DR. S: But then the burden sharpens. If black holes are this central, you do not get to treat them loosely.

DR. U: Correct. Their early appearance, their host relations, their jets, and their mergers all become part of the same cosmological test.

DR. S: Good. That is exactly where they should be.

The discussion had become stronger.
Not because the case was closed.
Because the architecture had acquired a physical center.

Genesis was no longer just a different word for origin.
It was a proposal about what kind of structure comes first, and why.

Dr. S pressed on.

DR. S: All right. Then say plainly what an island galaxy is.

DR. U: A finite, self-contained cosmic domain.

DR. S: More specifically.

DR. U: A galaxy in UFD is not just a luminous object floating inside one universal fluid history. It is an island universe — a bounded coherence domain with its own Genesis history, its own central organizing seed, its own extended pressure geometry, and its own enclosing boundary. That is why galaxies are not merely late objects inside cosmology. They are the primary cosmological units.

DR. S: So "island galaxy" is not metaphorical.

DR. U: No. It is structural.

DR. S: And bounded by what?

DR. U: By a galactic boundary.

DR. S: Define that carefully.

DR. U: A galactic boundary is the coherence-saturation shell associated with the Genesis history of an island galaxy. It is not just the visible edge of a spiral disk, and it is not just a bookkeeping surface for missing mass. It is a real geometric transition in the larger field structure.

DR. S: So this is a predicted structure.

DR. U: Yes. And a load-bearing one.

That was where the model became expensive in exactly the right way.
A weaker theory would have kept the boundary vague so it could do

whatever the argument needed later:
A stronger one made it carry weight immediately.

Dr. S pushed on the larger architecture now.

The Cosmic Crystal.

DR. S: Fine. But now I want to know where this is going. You are not just describing a local origin event. You are describing a whole cosmic architecture. So what is the larger picture?

DR. U: The larger picture is that the UEF does not remain shapeless. It relaxes toward ordered large-scale structure: a cosmic crystal.

DR. S: Ordered in what sense?

DR. U: In the sense that a saturated plenum tends toward a minimal-strain, lattice-like organization. In UFD, that late-time attractor is approximated by the term "Cosmic Crystal."

DR. S: That is a bold phrase.

DR. U: It should be heard as a physical one. The claim is not that the universe literally resembles a jewel. It is that a saturated medium may relax toward a large-scale, low-strain, cellular order, just as ordinary systems often settle into patterned minima rather than remaining shapeless.

DR. S: And the Genesis Event is part of that?

DR. U: More than part of it. The Genesis Event is the local birth of that architecture. It is where the deeper harmonic order first becomes a bounded substantive domain. You could say the Cosmic Crystal is the large-scale ordering tendency, and each Genesis Event is one local act by which that order precipitates into a finite world.

DR. S: So the Cosmic Garden grows inside the Cosmic Crystal.

DR. U: Exactly. The crystal gives the ordering tendency. The garden is the unfolding of many local births within it.

That improved the whole picture.
The alternative cosmology was no longer just "not the Big Bang."
It had become a positive story: harmonic order, local birth, central seed, black-hole organization, boundary formation, island development, and large-scale lattice-like order.

Dr. S then moved to the most obvious observational pressure point short of the microwave background.

The surprisingly early galaxies and black holes.

DR. S: All right. Then what does this buy you observationally before we get to the microwave sky?

DR. U: It buys continuous galaxy formation.

DR. S: Meaning galaxies can form at many epochs, not only in one early privileged era.

DR. U: Exactly. If Genesis Events are local rather than singular and universal, then one should expect young and mature galaxies across the observable timeline. One should also expect a spectrum of outcomes: failed seeds, intermediate systems, and massive outliers depending on local fuel conditions.

DR. S: Which means surprisingly mature galaxies at high redshift become less shocking.

DR. U: Yes. And very large black holes in small hosts become less shocking too. Those are not awkward exceptions in this picture. They are natural outcomes of local Genesis under different conditions.

DR. S: So the cosmic timeline becomes less like one synchronized history and more like many local histories unfolding inside one larger ordering field.

DR. U: Exactly.

That was one of the stronger moments in the exchange.
Not because it closed the case.
Because it gave the theory a clear and risky advantage: what looks "too early" or "too mature" in a one-history cosmology may look perfectly natural in a cosmology built from many local histories.

Dr. S then widened the frame again.

DR. S: Let me tell you what I think is attractive here. You are not merely replacing one creation image with another. You are trying to replace one kind of cosmological unit with another. In the standard picture, the basic unit is one expanding universe. In your picture, it is the bounded island galaxy born through Genesis and matured inside a larger ordered medium.

Dr. U: Yes.

Dr. S: That is a real shift.

Dr. U: It has to be.

Dr. S: And the danger is equally obvious. Once you do this, every downstream cosmological burden gets harder, not easier.

Dr. U: Of course.

Dr. S: Because now galaxies, supermassive black holes, dark-matter phenomenology, the microwave background, lensing, large-scale structure, and observer position all have to move together.

Dr. U: Exactly.
That was the center of gravity.
The alternative cosmology was no longer a prettier origin story.
It was a replacement architecture.

Dr. S was not merely asking whether Genesis sounded more physical than a singularity.
He was asking whether it could carry the same explanatory load without quietly borrowing back the standard picture.

Nothing about that was settled.
But the burden had become exact.

Dr. S: Let me try the summary test.

Dr. U: Go ahead.

Dr. S: On your view, cosmic origin is not fundamentally a single universal fireball expanding into thinner space. It begins through local Genesis Events: coherence quenches in the deepest field layer that project bounded UEF domains. Those domains mature as island galaxies organized around central vortex seeds, with supermassive black holes functioning as early and enduring coherence engines rather than merely late collapse products. Each island is bounded by a real galactic boundary, not just a visible edge. And the larger cosmic setting is not mainly stochastic diffusion in an expanding background, but the growth of a Cosmic Garden within the larger ordering tendency of a Cosmic Crystal.

Dr. U: That is fair.

Dr. S: The attraction is obvious.

DR. U: Yes.

DR. S: The danger is equally obvious.

DR. U: Yes.

DR. S: Good. Then the real dispute is finally clear. If black holes are early architectural features rather than late accidents, and if galaxies are bounded local worlds rather than debris in one universal expansion history, then the sky should retain some memory of those births. The next question is whether the background itself carries that memory.

DR. U: Exactly.

INTERLUDE

By the end of the exchange, the argument was no longer merely about whether the Big Bang was philosophically satisfying. It had become a dispute about what kind of unit the universe is built from.

The model's deepest shift was not simply from one origin story to another. It was from one basic cosmological unit to another

In the standard picture, the basic unit is one expanding cosmic history. In the UFD picture, the basic unit is the bounded island galaxy born through Genesis, organized around a central seed, structured by an early black-hole engine, enclosed by a real boundary, and embedded in a larger ordered plenum. That is more unified if it works. It is also much more exposed.

The black-hole discussion mattered because it gave the replacement cosmology a physical core. Genesis no longer looked like a vague alternative to singularity language. It became a proposal about how bounded worlds organize themselves from the start: central loading, central seed, central exhaust, central coherence.

The Cosmic Crystal mattered because it made the replacement cosmology deeper than a local origin story. Genesis was no longer only a way to avoid a singularity. It had become the local act by which the deeper ordering tendency of the medium precipitates into bounded worlds. The Cosmic Garden mattered because it turned one cosmic beginning into many local births, each with its own history and maturity.

And the next burden now came into view.

If galaxies are born as bounded local worlds, and if their central engines and boundaries are not secondary but original, then the larger medium should retain some trace of that structure. The microwave sky could no longer be treated as somebody else's problem. It had become the unavoidable test of whether the medium really remembers Genesis.

The gain was obvious.
So was the cost.

Temporary Verdict

Dr. U succeeded in making the alternative cosmology more physically legible. Genesis replaced primordial singularity, bounded island formation replaced one universal early diffusion, supermassive black holes became early organizing structures rather than awkward late arrivals, and the Cosmic Crystal / Cosmic Garden picture gave the model a deeper large-scale architecture rather than only a different beginning.

Dr. S succeeded in forcing the decisive pressure. A replacement cosmology does not earn its place by offering a more tangible origin story. It earns it only if Genesis, boundary structure, early SMBH formation, and the larger island-galaxy architecture can survive the same observational burdens that made the standard account powerful in the first place.

IF GALAXIES ARE BORN AS BOUNDED LOCAL WORLDS AND THE LARGER MEDIUM CARRIES A REAL MEMORY OF THAT BIRTH, THEN THE NEXT BURDEN IS OBVIOUS:

THE MICROWAVE SKY ITSELF.

Chapter 8

The Cosmic Microwave Background

Relic radiation, or boundary imprint?

Modern cosmology treats the cosmic microwave background, or CMB, as one of the central pillars of the standard picture of the universe. The story of how it was discovered is one of the most celebrated in twentieth-century physics. In 1965, Arno Penzias and Robert Wilson, working at Bell Labs, found a persistent microwave hum in every direction of the sky. It did not vary with time of day or season. They checked the instrument, the wiring, and every obvious local source. Nothing made it disappear. What they had found was a faint, nearly uniform glow of microwave radiation filling the observable sky at a temperature of about 2.7 degrees above absolute zero.

The interpretation came quickly. If the universe had once been in a hot, dense state and had expanded over billions of years, then radiation from that early phase should still remain, stretched by cosmic expansion into the microwave band and cooled almost to nothing. The newly discovered background radiation was therefore taken as one of the strongest confirmations of the hot Big Bang picture.

In the decades since, the CMB has only grown more important. COBE showed that the background has an almost perfectly thermal spectrum, just as one would expect from radiation once in equilibrium with hot matter. Later missions, especially WMAP and Planck, mapped tiny temperature variations across the sky with extraordinary precision. Those variations — only a few parts in a hundred thousand — are interpreted in the standard picture as the frozen imprint of sound waves in the hot early plasma, captured when the universe cooled enough for atoms to form and light to travel freely. From those variations, cosmologists have inferred a remarkably detailed picture of the age, composition, and large-scale geometry of the universe.

The strength of that picture is obvious. It gathers an enormous range of observations into one coherent historical story.

UFD, however, tells a different story. It does not treat the 2.7 Kelvin glow as leftover radiation from a primordial fireball. It treats it as the natural background temperature of the propagation field itself — the steady thermal hum of a real medium filling space. The tiny temperature variations across the sky are then given a different origin as well. Instead of being the frozen echo of sound waves in the early universe, they are read as patterns imprinted on the field by the boundary of our own bounded cosmic region — the same Genesis-based environment introduced in the previous chapter. On this view, the smooth background and the fine angular structure do not come from one event. They have distinct physical origins.

That is a very large departure from the standard account. It is also the model's central risk. So the real question is unavoidable: can a boundary-cavity picture carry any substantial fraction of the empirical weight that the standard interpretation of the CMB now carries? If it can, then the model has crossed a threshold very few alternatives ever reach.

TECHNICAL NOTE — THE BOUNDARY-CAVITY SECTOR

In UFD, the CMB TT spectrum is not treated as relic radiation from primordial plasma oscillations. It is treated as the response of an incoming global ULF field to a resonant boundary-cavity system associated with a galactic coherence-saturation shell. Within this picture, the first six TT peak positions are recovered quantitatively, the TT envelope has been developed to a semi-quantitative level, and the same phase structure gives rise to a preliminary TE extension.

Two structural constraints are already clear. First, a passive thin-slab boundary is not sufficient. Second, the shell is too thin to sustain its own internal resonance ladder. The boundary must therefore act by imposing effective, multipole-dependent conditions on a larger cavity response rather than by functioning as an independent resonator.

This sector has therefore moved beyond pure reinterpretation and into constrained quantitative structure. But it is not yet a full replacement for the standard CMB framework. Important burdens remain open, including a full transfer function, detailed peak-height structure, complete polarization amplitudes, a derived damping law, realistic observer geometry, and cross-consistency with lensing and related observables.

Dr. S came into the conversation with the confidence of a physicist standing on one of the strongest observational pillars in science.

DR. S: All right. You have taken on expansion, dark energy, and the Hubble story. Fine. Now we reach the point where cosmology stops being forgiving. What is the cosmic microwave background?

DR. U: A global field seen through a boundary.

DR. S: So not relic radiation from recombination.

DR. U: Not fundamentally, no.

DR. S: Then let me say at once what that means. You are not just challenging one interpretive habit. You are challenging one of the central empirical triumphs of modern cosmology.

DR. U: Yes.

DR. S: Good. Then give me the strong version.

DR. U: The strong version is that the background is not the afterglow of a primordial plasma becoming transparent. The 2.7 Kelvin monopole is the large-scale equilibrium state of the ULF itself. The angular structure is something else: a boundary-imprinted modulation associated with our own large-scale coherence environment.

DR. S: So the monopole and the anisotropy map do not share the same cause.

DR. U: Not on this view, no.

DR. S: That is already a major departure.

DR. U: It has to be.

That was the first real point of strain.
The standard framework gains enormous power by tying the whole microwave sky to one early-universe thermal story.
UFD answers by loosening that knot and assigning different parts of the observed sky to different physical causes.
Whether that is insight or evasion was exactly what Dr. S wanted to find out.

So he pressed where the standard case was strongest.

The spectrum itself.

A blackbody spectrum matters because it is one of the cleanest signatures of thermal equilibrium in all of physics. That is one reason the standard cosmological interpretation draws so much strength from it.

DR. S: Fine. Then let us begin with the hardest part. Forget the peaks for a moment. The blackbody spectrum by itself is already a huge success for the standard story. It is not just vaguely thermal. It is extraordinarily clean. Why should I think your alternative can even begin to account for that?

DR. U: Because the blackbody fact and the relic-plasma interpretation are not the same claim.

DR. S: Explain that carefully.

DR. U: A blackbody spectrum tells you that the radiation field is in, or very near, thermal equilibrium. It does not by itself tell you which physical system established that equilibrium. Standard cosmology says it was an early plasma. UFD says the equilibrium belongs instead to the large-scale ULF background.

DR. S: So your argument is not that the blackbody spectrum is unreal.

DR. U: Of course not. The spectrum is completely real. The question is what kind of medium is in equilibrium.

DR. S: And why should the ULF have a universal equilibrium temperature at all?

DR. U: Because if the ULF is a real medium spanning the large-scale cosmos, then it can have a large-scale equilibrium state just as any real medium can. The deeper claim is that the 2.7 Kelvin background is not a leftover flash from one ancient plasma phase. It is the present equilibrium glow of the light-bearing field itself.

DR. S: That is a very strong reinterpretation.

DR. U: It is. But notice what it buys. It separates two questions that standard cosmology fuses together: why the background is thermal, and why the map has the angular structure it has.

DR. S: So in your picture, the thermal smoothness is one problem and the patterned sky is another.

DR. U: Exactly.

That helped because it made the disagreement cleaner.
The issue was no longer whether the microwave background is thermal.
It was whether thermality forces one specific historical story.

Dr. S, predictably, pushed harder.

DR. S: All right. But standard cosmology does not just give me thermality in the abstract. It gives me a specific early-universe mechanism that makes a blackbody spectrum natural. What is your mechanism?

DR. U: Large-scale field equilibration.

DR. S: That is still too thin.

DR. U: Then take the more specific version. UFD treats the ULF as a real light-bearing medium, not as an empty stage through which radiation merely passes. If such a medium has structured, quantized modes and enough time to settle, then a universal thermal spectrum is not surprising. It is what a deeply equilibrated radiation field should look like.

DR. S: Why should the Planck form be natural in that medium rather than just some vague thermality?

The question carried more weight than its phrasing suggested. The Planck spectrum is one of the most famous shapes in physics: the exact curve describing how a perfectly thermalized system emits radiation across wavelength. The CMB is the cleanest Planck spectrum ever observed. Any reinterpretation has to explain not only why the radiation is thermal, but why it is thermal in this very specific, deeply quantized form.

DR. U: Because UFD does not treat quantization as a rule imposed from nowhere. It tries to root quantization in the medium itself. Stable excitations come in discrete circulation units rather than arbitrary fractions, so the field does not support just any possible mode structure. And because those modes belong to one continuous field, they need not remain isolated from one another. Through the constitutive dynamics of the medium, energy can redistribute across modes until the background approaches a common equilibrium distribution.

DR. S: So your claim is that the blackbody becomes physically intelligible once light is treated as excitations of a real quantized medium.

DR. U: Exactly. On that picture, the extraordinary smoothness of the spectrum is no longer merely a historical leftover. It becomes the equilibrium signature of the medium itself.

DR. S: Have you fully derived the exact Planck spectrum from the constitutive dynamics of the ULF?

DR. U: No. That would be too strong. The present claim is more disciplined than that.

DR. S: Say it carefully, then.

DR. U: UFD offers a deeper physical rationale for why such a blackbody can exist without a primordial fireball: a real medium, quantized modes, and long-term re-equilibration. But a closed first-principles derivation of the exact monopole spectrum remains part of the burden.

DR. S: And if a spectral distortion were found?

DR. U: Then the interpretive question changes. In the standard view, the distortion preserves a memory of early-universe energy injection. In UFD, one would first ask what later process disturbed the ULF faster than it could relax — local energetic events, boundary dynamics, or other departures from equilibrium.

That concession mattered.
A weaker theory would have dodged the blackbody problem with rhetoric. A stronger one admitted exactly where the reinterpretation is physically legible and where it still owes mathematical closure.

Only then did Dr. S move to the peaks.

When astronomers map the temperature variations of the CMB across the sky, they find a series of preferred angular scales. Plot the variation power against angular scale and the result is the famous series of bumps known as the acoustic peaks. In standard cosmology, those peaks are tied to sound waves in the primordial plasma and to the sound horizon at recombination.

UFD was replacing that origin story with another one. Instead of peaks shaped by primordial plasma oscillations, it reads them as patterns shaped by the boundary of our own bounded region.

DR. S: Fine. Then let us take the peaks. In the standard view, the acoustic peaks come from oscillations in the primordial plasma, filtered through recombination and later propagation. What are they here?

DR. U: Harmonic structure tied to a boundary coherence scale.

DR. S: Meaning literally a galactic boundary mode.

DR. U: In the broad sense, yes. The smooth field that fills space comes to us through the boundary of our own region — the edge that formed during our galaxy's founding event. That boundary is not featureless. It has a characteristic size on which patterns naturally lock into place, the way ripples on a drumhead settle into preferred shapes determined by the rim. The spacing of the peaks reflects that size. It is not the leftover signature of sound waves in an early plasma. It is the geometric fingerprint of the boundary itself.

DR. S: So the peak spacing is not a sound horizon.

DR. U: Not in UFD. It reflects a resonant boundary scale.

DR. S: And you now mean that more strongly than before. Earlier you were offering a structural reinterpretation. You are now claiming that the boundary picture produces specific numbers.

DR. U: Yes. This is no longer only a verbal reassignment. The peak-position sector can be written in forward form. The first six peaks in the TT spectrum can be reproduced by a resonant boundary response.

DR. S: That is stronger than a toy analogy.

DR. U: It is. But only in one sector. The point is not that the whole microwave program is finished. The point is that the boundary picture reaches a real quantitative burden and survives it at the level of peak positions.

DR. S: All right. Then tell me what has changed. Where has the boundary picture actually advanced?

DR. U: Two things. First, the boundary itself can no longer be treated vaguely. The simplest version of the model — a passive shell with no real internal role — fails decisively. Second, the model has moved beyond peak spacing alone. It now reproduces the broader TT envelope in a rough but recognizable way.

DR. S: Semi-quantitatively is doing important work there.

DR. U: It is. I am not claiming the boundary picture can yet match the full precision of the standard transfer calculation across every feature of the spectrum. I am saying it has moved beyond fitting one set of peak positions to also capturing the broader envelope across scales.

DR. S: And the shell itself? What exactly has changed there?

DR. U: We used to leave open the possibility that the shell itself was somehow ringing internally and producing the pattern. That picture has been ruled out. A passive shell fails, and a shell treated as an internal ladder of resonances fails as well.

DR. S: Go on.

DR. U: The right picture is not that the shell rings by itself. The right picture is that the whole region inside the boundary supports the pattern — like a drumhead rather than just a rim — while the shell sets the edge conditions. The shell shapes the modes. It is not the source of them.

DR. S: That is actually an important clarification.

DR. U: Exactly. It narrows what the model has to do. The shell cannot remain a vague resonance surface we hope will somehow handle everything. The model now has to live or die by a real cavity-and-boundary response that produces the right patterns at the right scales.

That was one of the model's biggest gains.
UFD was not saying, *the peaks are fake.*
It was saying, *the peaks are real, but they come from a different physical structure, and that structure cannot be arbitrarily simple.*
That is a genuine alternative claim.
It is also expensive in exactly the right way.

So Dr. S pressed on whether the framework had anything more than labels.

DR. S: Here is my main worry. This kind of thing can sound ingenious because every standard feature gets a new label. Peak spacing becomes boundary scale. Damping becomes impedance disorder. Fine. But relabeling is cheap. Do you actually have a forward theory?

DR. U: Not a completed one. But there is more than relabeling.

DR. S: Go on.

DR. U: The present model does real work. It does not yet give a finished transfer code, and it is not being presented as one. But once you separate the monopole from the anisotropy mechanism, the boundary-response picture can already do three things. It can recover the TT peak positions quantitatively. It can reproduce the TT envelope semi-quantitatively. And it can carry a preliminary structural extension into the TE sector.

DR. S: Explain that last phrase carefully.

DR. U: The claim is modest but real. The same cavity-phase scaffold that organizes the TT structure also yields the sign alternation in the TE sector.

DR. S: Sign alternation, not amplitude closure.

DR. U: Exactly. The phasing is already linked to the same boundary-cavity logic. But the detailed TE amplitude envelope, the EE sector, and the full tensor response of the boundary are still open.

DR. S: Better. Because that is exactly the kind of maturity distinction that keeps an alternative from inflating itself.

DR. U: It has to. Otherwise, the model weakens itself.

That made the conversation stronger.
A weaker theory would have pretended closure it did not have.
A stronger one marked the gain and the remaining burden separately.

Dr. S respected that enough to move on to polarization more directly.

Polarization is the directional structure of the microwave signal, and in standard cosmology it carries a great deal of information about early-universe conditions. That is why any alternative has to explain it eventually rather than treat it as optional.

DR. S: All right. Then what about polarization in the fuller sense? Standard cosmology treats the E- and B-mode structure as part of the same early-universe story. What happens here?

DR. U: The same division of labor applies. Polarization is not a separate mystery. It is another response channel of the same boundary-imprint system.

DR. S: More concretely.

DR. U: Compressional response from the Genesis imprint would feed one polarization pattern. Shear or rotational response would feed another. The important point is that TT and EE do not need separate fundamental scales here. They can arise as different response channels of one underlying boundary coherence structure.

DR. S: But not yet quantitatively.

DR. U: Not at the amplitude level, no. The TE sign structure has begun to emerge from the same phase scaffold, which is already something. But the full polarization sector is still open.

DR. S: Good. That is clearer.

That was useful.
It kept the model from becoming a grab bag of disconnected reinterpretations.
Monopole, peaks, damping, and polarization were all being gathered into one medium-and-boundary picture.

Dr. S then moved where such pictures usually become either serious or vague.

Lensing.

Lensing matters because it tests the story through the later gravitational universe, not only through the background signal itself. If the microwave sky has really been deflected by intervening structure, then any alternative must preserve that long-path optical history in a self-consistent order.

DR. S: All right. Then let us talk about lensing, because that is where stories start losing their freedom. What happens to the observed lensing signal in your picture?

DR. U: It stays, but the optical order changes.

DR. S: Go on.

DR. U: In UFD, the local boundary is not the terminal source of the observed structure. It is a resonant imprint surface for an incoming global ULF field. So the sequence is: global propagation first, then cosmological lensing, then local boundary imprint, then observer.

DR. S: That had better be the sequence.

DR. U: It has to be. If the boundary were the whole source, the available path length would be far too small to account for the observed lensing.

DR. S: So the same large-scale pressure geometry you have already invoked elsewhere now has to reproduce the lensing variance here as well.

DR. U: Exactly. If the UEF-based lensing picture cannot reproduce the observed microwave-background lensing signal, then this part of the model fails.

DR. S: So the CMB sector is now stronger in one way and still exposed in another.

DR. U: Yes. The TT and preliminary TE structure have become more constrained. The full lensing cross-consistency still remains a major open burden.

That was the conversation's first true hard edge.
Not whether the reinterpretation was imaginable.
Whether it stayed self-consistent when two different burdens met each other.
If galactic lensing and background lensing are both supposed to arise from one pressure-geometry ontology, then UFD becomes vulnerable in exactly the right place.

Dr. S pushed again.

DR. S: There is another obvious problem. We are not at the center of the galaxy.

DR. U: Correct.

The objection was sharper than its phrasing suggested. If UFD's reading of the CMB depends on our galaxy's boundary, then our position inside the galaxy matters. We are not at the galactic center. A naive local-boundary model would predict obvious asymmetries in the sky. But the microwave sky is one of the most isotropic signals in observational physics. Any model that places its structure within our own galactic environment has to clear that bar.

DR. S: Then why is the microwave sky so isotropic?

DR. U: Because the isotropic monopole is not being assigned to the shell in the first place.

DR. S: Go on.

DR. U: That is exactly why the monopole-anisotropy separation matters so much. The smooth 2.7 Kelvin background belongs to the global ULF equilibrium, not to local shell brightness. The boundary acts only on the perturbation sector.

DR. S: So our offset does not automatically destroy the picture.

DR. U: Not at high multipoles. For the shorter-wavelength angular structure, the distortion from observer offset falls rapidly. The serious burden lies at low multipoles, where mode coupling becomes much more important.

DR. S: Which means the off-center problem has not disappeared. It has just been reassigned to the low-order anomaly sector.

DR. U: Exactly. And that is where a realistic treatment would require boundary asphericity and orientation, not just the simplest spherical shell.

That mattered because it kept the picture honest.
A weaker theory would have hidden the observer-position problem behind vague geometry.
A stronger one admitted that it turns the low-order sky into a sharp constraint.

Dr. S then widened the frame one final time.

DR. S: Let me tell you where I think the real issue is now. Standard cosmology buys a lot by tying the blackbody spectrum, the peaks, the polarization, and the lensing into one early thermal history. Your framework buys something else. It breaks that one story apart and replaces it with medium equilibrium plus boundary response plus propagation.

DR. U: Yes.

DR. S: The attraction is obvious. It reduces the need for an elaborate primordial script.

DR. U: Yes.

DR. S: The danger is equally obvious. It has to do more geometric work in the present structure of the universe.

DR. U: Exactly.

DR. S: And right now, if I am being fair, this is still one of the places where your framework looks most ambitious and least complete.

DR. U: Fair enough. But the model is stronger than that sentence alone would suggest. One part of it has crossed from conceptual plausibility into quantitative exposure. Another has crossed into semi-quantitative development. A third has reached a preliminary structural extension. The rest remains unfinished.

DR. S: Good. That is the right way to say it.

That was the right line.
Not because it resolved anything.
Because it named the situation honestly.

In some domains, a theory's strength is that it already fits.
In others, its strength is that it defines a sharper replacement problem than anyone else has defined.
This was plainly one of those.

Dr. S did not respond immediately. He turned the summary over once more before speaking, the way a referee handles a paper whose ambition has outrun its current evidence.

DR. S: What you have done is not small. You have taken the single most precisely measured signal in cosmology — the spectrum that effectively founded twentieth-century cosmological precision — and reassigned almost every feature of it to a different physical cause. The blackbody belongs to a medium rather than to a fireball. The peak spacing belongs to a boundary rather than to a sound horizon. The envelope belongs to a cavity response rather than to acoustic transfer. The polarization is starting to belong to the same scaffold. That is not a refinement. That is a wholesale reinterpretation of the observable.

DR. U: I am aware.

DR. S: Then let me say what I find genuinely interesting about it, and what worries me. What is interesting is that the picture is internally coherent in a way that earlier alternatives to the standard CMB story were not. Tired-light models failed because they could not produce time dilation from the same mechanism as redshift. Most steady-state revisions failed because they could not produce a thermal spectrum at all. You are not making either of those mistakes. The thermal spectrum is built into the medium from the start. The peak structure now carries quantitative weight. The envelope is moving in the right direction. As a structural proposal, this is a more disciplined alternative than the field has seen in a long time.

DR. U: And the worry?

DR. S: The worry is that the standard story is not just internally coherent. It is the most over-constrained model in cosmology. Every feature you have just reassigned — the spectrum, the peaks, the envelope, the polarization, the lensing — was independently predicted before it was measured by one early-universe history with a small number of free parameters. The agreement between those predictions and the data is the most precise observational triumph in modern physical science. To reinterpret all of that within a different framework, you have to reproduce not just the qualitative features but the same numerical constraints, and you have to do it without quietly importing the early-universe story through the back door. That is not a burden you can discharge with a coherent picture. It is a burden you can only discharge with a completed calculation that survives every test the standard model has already survived.

DR. U: I know.

DR. S: Good. Then we both know what is at stake. You are not asking me to prefer your picture today. You are asking me to take it seriously enough to test it. Those are different things, and the second is the harder ask, because taking it seriously means doing the work to break it — and most of the people who could do that work have no incentive to.

DR. U: That is the structural problem.

DR. S: It is. But it is not your problem to solve in this conversation. Your problem in this conversation was to show that the picture has reached the point where the test would be worth doing. I think you have shown that. I do not yet think you have shown more.

DR. U: That is the right verdict.

DR. S: Then let us leave it there.

INTERLUDE

By the end of the exchange, the microwave sky no longer looked like a simple photograph of a hot beginning.

At least in the UFD picture, it had become a layered signal: a global equilibrium field seen through local structure.

That was the gain.

113

The blackbody background and the patterned map no longer had to share one cause. The thermal smoothness could belong to the large-scale ULF itself, while the anisotropies could belong to boundary imprint, propagation, and lensing through a real medium. The model's gain is now stronger than pure reinterpretation. The TT peak-position sector has crossed into quantitative exposure. The TT envelope has moved into semi-quantitative development. The preliminary TE sign structure matters because it means the same phase scaffold is beginning to organize more than one observable channel. The failure of the passive shell picture matters even more because it means the boundary cannot remain vague. If the model is right, the shell must act as a real thin-shell coherence interface imposing effective boundary conditions on a larger cavity response.

That does not finish the job.

But it does make the job physically real.

That was also the cost.

A reinterpretation of the microwave sky earns nothing by being imaginative. It earns its place only if it can become a full forward theory: one that accounts for the blackbody spectrum, angular structure, polarization amplitudes, lensing, and observer geometry without quietly borrowing back the very relic-emission story it claims to replace.

The gain is clearer now.
So is the unfinished burden.
And that honesty makes the model stronger rather than weaker.

TEMPORARY VERDICT

Dr. U succeeded in making the alternative microwave-background picture physically serious. The blackbody spectrum received a real medium-based reinterpretation rather than a denial, the anisotropy structure became a boundary-cavity response problem, the TT peak-position sector gave the framework a genuine quantitative foothold, the TT envelope reached semi-quantitative development, and the TE sign structure began to follow from the same phase scaffold. Most importantly, the model did more than attach new language to familiar data. It imposed real constitutive discipline on itself: the passive shell picture failed, the shell was forced into the thin-shell regime, and the boundary could no longer be treated as decorative. It had to become a real active coherence interface.

Dr. S succeeded in forcing the decisive pressure. A replacement cosmology does not earn its place by offering a more tangible sky story. It earns it only if the replacement can carry the same observational weight with real forward consistency. That means not only peak positions, but full transfer structure, polarization amplitudes, realistic observer geometry, and lensing cross-consistency.

IF THE MICROWAVE SKY CAN BE RE-READ WITHOUT A UNIVERSAL FIREBALL, THEN THE NEXT BURDEN IS BROADER THAN COSMOLOGY ALONE. THE QUESTION BECOMES WHETHER THE SAME MEDIUM-BASED ARCHITECTURE CAN ALSO MAKE LIGHT, RELATIVITY, THE WEAK INTERACTION, ELECTROMAGNETISM, AND QUANTUM BEHAVIOR PHYSICALLY INTELLIGIBLE AS PARTS OF ONE CONTINUOUS WORLD-PICTURE.

CHAPTER 9

Light and Physical Relativity

A real light field, and the strange rules that follow from it

Modern physics inherited two great puzzles about light. The first was that light behaves like a wave. It interferes, diffracts, and spreads through space in orderly patterns. In the 1860s, James Clerk Maxwell captured this behavior with extraordinary precision by unifying electricity, magnetism, and light into one electromagnetic theory. Maxwell's equations implied that light propagates as a wave at a definite speed set by the electric and magnetic properties of empty space. That was one of the decisive triumphs of nineteenth-century physics. Light was a wave, and the wave was electromagnetic.

But waves had always been understood to require a medium. Sound moves through air. Water waves move through water. If light was a wave, it seemed natural to assume that it too moved through something. Physicists gave that hypothetical medium a name: the luminiferous ether. Then came one of the most consequential null results in science. In 1887, Michelson and Morley looked for Earth's motion through the ether by comparing the speed of light in different directions. They found nothing. No matter how the apparatus was oriented, the speed of light came out the same. The medium that the wave was supposed to move through could not be detected.

Then came the second puzzle. By the early twentieth century, light was also behaving like a particle. It arrived in discrete packets, triggered the photoelectric effect one electron at a time, and refused to behave like a smooth classical disturbance whenever it exchanged energy with matter. Light seemed to be both wave and particle at once.

A third puzzle sat behind both. The speed of light appeared to be the same for all observers, no matter how they moved. A flashlight on a moving train and a flashlight at rest both produced light with the same measured speed. There was no ordinary way to add or subtract the source's motion from the speed of the light itself. And matter, for its part, resisted acceleration in a way physics still usually treated as primitive rather than explained. Inertia remained a basic fact.

Einstein's revolutions resolved these problems mathematically, but by elevating certain empirical facts into foundational principles. The constancy of the speed of light became a postulate of special relativity. Quantized light-matter exchange became part of the new quantum framework. Inertia remained fundamental. The mathematics worked beautifully, and the predictions of relativity and quantum theory have survived more than a century of experiment. But the deeper physical questions were not answered so much as absorbed into the formalism. Why is the speed of light invariant? Why does matter resist acceleration? Why does light behave as both wave and particle? Standard physics gives extraordinarily successful rules. It is less settled on deeper physical causes.

This conversation matters because light sits at the crossroads of modern physics. Wave behavior, quantum discreteness, and relativity all meet there. To reinterpret light is therefore to say something new about Maxwell, Einstein, and quantum theory at once. It is one of the most exposed places any framework can choose to engage.

UFD ties the three puzzles together. Light is a real propagating excitation of the Universal Light Field. Its wave behavior is literal because it really is a wave in a medium. Its particle-like behavior appears when that wave exchanges energy with matter in quantized, resonant ways, because matter itself is built from structured field excitations and can only absorb or emit energy in discrete units. And the strange rules of relativity arise because clocks, rods, and particles are not external observers of the light-bearing medium. They are built from it. Motion through the ULF changes the physical state of the moving system. Internal processes slow, geometry compresses along the direction of motion, and inertial resistance rises because the structure must carry an increasing distortion burden in the field.

On this view, light and relativity are not separate mysteries. They are two sides of the same medium-based reality. The constancy of the speed of light becomes a derived consequence of how the medium responds to motion through it. Inertia becomes another consequence of the same response. The photon becomes a real excitation of a real field rather than a merely axiomatic ingredient of the equations. None of those are small claims. Each one reopens a question mainstream physics usually treats as closed.

By the time they reached this conversation, Dr. S already knew the general shape of the dispute. Dr. U wanted physical causes where modern physics often stopped with successful rules. Dr. S did not object to deeper causes in principle. He objected to them when they made the world sound more vivid without becoming more exact. So he began where the whole issue begins.

Light.

DR. S: Fine. Tell me what light is.

DR. U: A real wave in the ULF.

DR. S: Not a disturbance in emptiness.

DR. U: Not fundamentally, no.

DR. S: Then let me state the standard view cleanly. Classical physics made light a wave. Quantum theory made it a photon. Relativity made its speed invariant. The mathematics works magnificently even if the physical picture remains strange. Why is that not enough?

DR. U: Because the three mysteries remain disconnected. Why should a wave act like a particle at interaction? Why should its speed be fixed for all observers? Why should matter resist acceleration more and more as velocity rises? In UFD, those are not separate questions. They are different expressions of one field architecture.

DR. S: That is the ambition. Now give me the mechanism.

DR. U: Light is a propagating ULF wave packet. It is not a little billiard ball with wave-like habits. It is a real wave packet in a real field.

That immediately clarified the terms of the dispute.
The question was no longer whether light's mathematics works.
It was whether the mathematics describes something physically intelligible.

Dr. S pushed at once to the old quantum puzzle.

Wave-particle duality.

This was one of the oldest conceptual fractures in modern physics. The experiments were clear. What remained unclear was what kind of thing could behave as both a wave in transit and a discrete event at detection.

DR. S: Fine. Then what becomes of wave-particle duality?

DR. U: It stops being a duality in the deepest sense. Light is a wave in transit and looks particle-like only in quantized interaction.

DR. S: More specifically.

DR. U: Think of the photoelectric effect. In standard quantum theory, light arrives as photons and kicks out electrons one by one. UFD keeps the discreteness of the event, but the picture changes. The incoming ULF wave packet interacts with the resonant structure of the material. If the packet carries enough energy to meet the local binding threshold, the whole packet is absorbed in one resonant event and one electron is released. That is why frequency sets the energy of the ejected electron and intensity sets how many events occur.

DR. S: So the discreteness belongs to the interaction, not to a tiny pellet flying through empty space.

DR. U: Exactly.

DR. S: And the double slit?

DR. U: A wave in action. The ULF packet passes through both available paths as a real extended excitation and interferes with itself because it really is a wave. The localized "particle" aspect appears only when the field couples resonantly to a detector.

DR. S: So your claim is that the paradox comes from trying to force one process into two incompatible ontologies.

Dr. U: Yes. In transit it is a wave. In quantized interaction it behaves discretely. That is one picture, not two.

That was one of the cleanest moves available to UFD.
It did not try to erase the data.
It tried to say what kind of world would make the data feel less contradictory.

Dr. S respected that enough to push to the next familiar principle.

Least action.

The principle of least action is one of the deepest organizing rules in modern physics. It works beautifully. What it does not obviously provide is a physical reason for why nature should favor one path out of many possible ones.

Dr. S: All right. Then what about the principle of least action? Standard physics treats it as one of the deepest mathematical principles in science. What does your medium buy you there?

Dr. U: A cause.

Dr. S: Go on.

Dr. U: In standard theory, least action is often treated as a powerful rule that somehow knows the right path in advance. In UFD, a real wave in a structured field naturally follows the path of least resistance. The field has density, elasticity, and local structure. The favored path is simply the physically easiest one for propagation through that medium.

Dr. S: So the mathematics survives, but it becomes the summary of a physical tendency.

Dr. U: Exactly.

That mattered.
A weaker reinterpretation would merely rename the principle.
A stronger one gave it a physical reason.

Dr. S then moved to the hardest part of the conversation.

Not light.

The speed of light.

Dr. S: Fine. Here is the real pressure point. Once you say there is a real propagation medium, you seem to invite a preferred frame. But the whole point of relativity is that no inertial observer detects one through the speed of light. So how do you avoid walking straight back into the ether problem?

DR. U: By making the ULF the physical basis of the invariance rather than its violation.

DR. S: Say that more carefully.

DR. U: Light speed is invariant because light, matter, rods, and clocks all belong to the same field architecture. You are never comparing a signal in the field to instruments outside the field. You are comparing one field expression to another. So the measured invariance survives because the measuring apparatus co-deforms with the propagation conditions.

DR. S: So the medium does not ruin the invariance because the observer is built out of the same medium.

DR. U: Exactly.

DR. S: That is a strong move.

DR. U: It has to be. Otherwise, the medium picture dies immediately.

That was the hinge of the whole discussion.
If the ULF were just an old-fashioned ether, the model would collapse right there.
The only way forward was to say that the signal and the measuring system are both field-made and therefore change together.

Dr. S heard that clearly and pushed to the first relativistic effect.

Time dilation.

Time dilation is one of the strangest and best-confirmed predictions of special relativity. Moving clocks really do run slower. The question is not whether the effect is real. It is what kind of physical story, if any, sits beneath the transformation law.

DR. S: Fine. Then take time dilation. What is it, physically?

DR. U: A slowing of the moving system's internal workings, caused by the strain of pushing through the field it lives in.

DR. S: More specifically.

DR. U: When an object sits at rest in its local field environment, its internal processes run at their natural rate. But when it moves quickly through the field, it has to restructure the surrounding medium as it goes. That takes real energy. The object now carries a distortion halo — a kind of bow wave in the field. Some of the system's energy is tied up in sustaining that disturbance. There is less left over for the internal processes that make the object tick. Those processes slow down.

DR. S: So the clock really runs slower.

DR. U: Yes. Not because time itself becomes magical, but because the physical processes that make the clock a clock are slowed by the burden of moving through the field.

DR. S: And you think that gives a causal account where standard relativity gives only a transformation rule.

DR. U: Exactly. Special relativity tells you what happens. UFD tries to tell you why it happens.

That made the disagreement exact.
Standard relativity says how clocks transform.
UFD tries to say what physically changes in the clock.

Dr. S moved immediately to the next effect.

Length contraction.

Length contraction is the spatial twin of time dilation. Standard relativity describes it with perfect success. UFD tries to make it physically legible.

DR. S: All right. Then what is length contraction?

DR. U: A real compression of the moving object into the easiest shape it can take while pushing through the field.

DR. S: That is stronger than the usual story.

DR. U: It has to be. When an object moves quickly through the field, it cannot keep its rest shape unchanged. The distortion halo is not free. It costs energy to maintain, and that cost depends on geometry. A long object moving through the medium displaces more field than a shorter one aligned the same way. So the system settles into the lower-cost transport shape.

DR. S: "Easiest" meaning lower field cost.

DR. U: Exactly. The same way a swimmer streamlines to move through water more efficiently, a moving object in the field deforms into the most economical shape available. The difference is that here the deformation is automatic.

DR. S: So the ruler really shortens.

DR. U: Yes. Not because an observer merely chooses coordinates differently, but because the object's actual supported geometry has changed under motion through the field.

That sharpened the model in exactly the right way.
A weaker reinterpretation would have said: relativity stays the same, but imagine it more vividly.
A stronger reinterpretation said: the effects are physically real because the medium is physically real.

DR. S: Wait. Before we go further, I need you to answer the one question that will determine whether any of this survives. Does UFD recover the Lorentz transformations exactly, or only approximately?

DR. U: Exactly, in the regime where they have been tested.

DR. S: Say that more carefully.

What Dr. S was asking for was the most important commitment Dr. U could make. Lorentz invariance is not a side issue. It is the formal heart of special relativity. Any medium theory that fails to recover it with full tested precision is dead on arrival.

DR. U: Lorentz invariance emerges as an exact symmetry of the field in the regime where standard physics has tested it. The reason is structural. Every measuring instrument we could ever build is itself an excitation of the same medium. The clocks, the rulers, the detectors, the observers — none of them stand outside the field. They are all built from it. When the field is perturbed by motion, the instruments built from it deform along with it. No internal measurement can detect the field's underlying state because there is nothing outside the field to compare it against. The symmetry is not imposed. It is automatic.

DR. S: So the invariance is exact by construction, not approximate by luck.

DR. U: Exact at the level of all present tests. The framework becomes distinguishable from standard relativity only if the field carries residual large-scale structure that might show up in ultra-precision measurements not yet performed.

DR. S: And if those effects never appear?

DR. U: Then the medium picture remains empirically equivalent to standard relativity in the tested regime, and its gain is explanatory. But it still risks subtle higher-order departures — for example, small asymmetries in time dilation between observers in genuinely different field environments — if the larger structure of the medium is real and detectable.

DR. S: That is an honest answer.

That mattered more than it might have seemed.
A weaker framework would have waved away the Lorentz question.
A stronger one said plainly: the invariance is recovered exactly in the tested regime, and the two frameworks become distinguishable only through small higher-order effects.

Dr. S immediately pushed where that strength became risky.

Relativistic inertia.

Inertia is the property that makes objects resist acceleration. Special relativity adds a remarkable twist: the faster an object moves, the harder it becomes to accelerate further. The effect is real. The question is where the extra resistance comes from.

DR. S: All right. Then inertia. Standard physics still usually treats resistance to acceleration as a primitive fact. What is it here?

DR. U: The energy cost of dragging a coherent object through the field.

DR. S: More specifically.

DR. U: Inertia in UFD is the work required to push a stable structure through the medium it lives in. The bow wave I described in the time-dilation discussion is the same picture. As velocity rises, the distortion halo grows. More energy has to be tied up in sustaining that disturbance. That is what makes further acceleration harder. The object is not just being moved. It is being moved together with the growing disturbed region of field around it.

DR. S: So the increase in inertia is not mysterious.

DR. U: Exactly. It is the cost of moving a coherent structure while preserving its identity in a field that resists rapid restructuring.

DR. S: And what about the old idea of relativistic mass?

DR. U: I keep the physical content and worry less about the label. What matters is that the extra energy is not "inside the particle" in some crude sense. It is the real energy of the enlarged distortion field the object must now sustain.

DR. S: So rest mass itself?

DR. U: The footprint of a stable structure in the field when it is not moving. Even at rest, the object has to maintain its own existence. The energy required to sustain that structure is what we measure as rest mass. Motion increases the burden without changing what kind of object it is. Rest mass is the cost of being. Relativistic increase is the added cost of moving.

The gain here was unification.
If the same field that carries light also carries the distortion burden that slows clocks, contracts lengths, and resists acceleration, then several standard "givens" begin to collapse into one mechanism.
That was one of the model's strongest moves.
Light, inertia, time dilation, and relativistic mass were no longer different topics.
They had been drawn into one common field story.

Dr. S recognized that immediately.
Which was why he moved to the deeper philosophical issue rather than the next formula.

DR. S: Let me ask the deeper question. Why do you care so much about making relativity causal? Standard relativity is one of the most successful frameworks ever devised. Why not leave it alone?

DR. U: Because it describes with extraordinary power what moving systems do, but it remains thin about why they do it. Time slows, lengths contract, inertia rises, c stays fixed — all true. But the standard picture asks you to accept those as features of spacetime structure. UFD asks whether they are the visible behavior of something physically happening in a field.

DR. S: So for you, the issue is not prediction. It is explanatory depth.

DR. U: Yes. And explanatory unification. The same field that carries light also carries the burden of inertia and the mechanism of relativistic distortion. That is better than having separate mysteries line up by formal coincidence.

DR. S: Unless the better story turns out to be false.

DR. U: Of course.

That was the center of the exchange.
Dr. S trusted the existing formal framework enough not to demand a
deeper story.
Dr. U distrusted stopping there, especially when the same field might unify
light, inertia, and relativistic distortion.
Neither position was absurd.
But the target had become exact.

So Dr. S gave the model its final test.

DR. S: Let me try the summary test.

DR. U: Go ahead.

DR. S: On your view, light is a real propagating excitation of the ULF, not a
disturbance in emptiness. Electric and magnetic behavior belong to one
medium. A photon is a self-sustaining ULF wave packet whose particle-like
character appears in quantized interaction rather than in transit. The speed
of light is the characteristic propagation speed of the medium, not a miracle
built into nothingness. Least action becomes ordinary path-of-least-
resistance behavior in a real field. And special relativity is not a purely
kinematic truth about empty spacetime. It is the observable behavior of
coherent structures moving through the ULF. Time dilation is the slowing
of internal resonance under a growing distortion burden. Length
contraction is real compression into a lower-cost transport geometry.
Increased inertia comes from the energy cost of carrying that distortion
halo. And the constancy of measured light speed survives because rods,
clocks, and signals are all expressions of the same field architecture rather
than independent things compared from outside it.

DR. U: That is fair.

DR. S: The attraction is obvious.

DR. U: Yes.

DR. S: The danger is also obvious.

DR. U: Yes.

DR. S: Good. Then the burden is finally clear.

By the end of the exchange, light and relativity no longer looked like separate achievements of twentieth-century physics that just happened to fit together.

At least in the UFD picture, they had become two aspects of one field-based reality.

That was the gain.

Light was no longer a strange hybrid of wave and particle by decree. It was a real wave packet whose discreteness appeared in resonant interaction. Least action was no longer a rule hovering above physics. It was the natural behavior of a wave moving through a structured medium. And relativity was no longer a purely abstract geometry of empty spacetime. It was the material symmetry of the very field that carries light.

That was also the risk.

A medium-based account of light and relativity earns nothing merely by sounding more intuitive. It earns its place only if it preserves the full empirical success of wave optics, quantum interaction, and Lorentz invariance while exposing itself honestly wherever a deeper medium might leave detectable traces — including subtle anisotropy and asymmetrical time-dilation effects if the higher-order structure of the plenum can eventually be probed.

TEMPORARY VERDICT

Dr. U succeeded in making light and relativity physically legible as one problem rather than two. Wave-particle behavior, light-speed invariance, time dilation, length contraction, and relativistic inertia were all drawn into one field-based mechanism.

Dr. S succeeded in holding the correct line of pressure: a causal medium story earns its place only if it preserves the observed symmetry with the same discipline as standard relativity, while exposing itself honestly where it claims more.

IF LIGHT AND RELATIVITY BOTH EMERGE FROM THE COMMUNICATIVE FIELD, THEN THE NEXT QUESTION IS HOW THE ELECTROWEAK WORLD LOOKS ONCE PARTICLE IDENTITY ITSELF BECOMES GEOMETRIC.

CHAPTER 10

The Weak Force and the Composite Neutron

Figure-eight collapse and electroweak change

The weak interaction has one of the strangest histories in modern physics. It was discovered before anyone really knew what it was. Radioactivity first appeared as an invisible leakage from unstable matter. Beta decay, in particular, quickly became a crisis, because the emitted electrons did not come out with one fixed energy. Something seemed to be missing from the bookkeeping. That puzzle led Pauli to propose the neutrino, and led Fermi to formulate a theory of beta decay that introduced a new fundamental interaction: the weak force.

The story grew stranger. In 1956, Chien-Shiung Wu and her collaborators showed that weak processes violate parity: they distinguish left from right in a way no other known force had done. Then, in the 1960s and 1970s, Glashow, Salam, and Weinberg folded the weak interaction together with electromagnetism into the electroweak theory, and the later discovery of the W and Z bosons made that framework one of the great precision triumphs of modern particle physics. The strength of the standard picture is obvious. It organizes a huge body of data within one disciplined mathematical structure, even if the underlying ontology remains abstract.

That is why this chapter is expensive. The weak interaction is not a peripheral corner of physics. It governs radioactive transformation, solar nuclear processes, neutrino-linked phenomena, and a large body of precision collider data. Any replacement picture has to face not only beta decay, but parity violation, the W and Z sector, the weak coupling scale, and the observed lifetimes of unstable systems. The bar is extremely high.

Unified Field Dynamics begins somewhere else. In UFD, the weak interaction is not treated as fundamental in that sense. It is treated as the visible signature of a deeper structural instability in matter. The proton is the ground-state heavy structure: a low-strain trefoil-like vortex in the dense field. The neutron is not a second equally fundamental heavy-matter endpoint, but a metastable composite formed under extreme Genesis conditions. At the broadest intuitive level, one can picture it as a proton-like source bound to an oppositely chiral sink-type heavy mode. More carefully stated, it is a pressure-locked figure-eight-like composite in the UEF.

Beta decay is then re-read not as a primitive weak event, but as a geometric relaxation cascade. The composite fails. The proton survives as the lower-strain endpoint. The electron recoheres in the ULF as the natural toroidal descendant of the shed circulation. And the antineutrino is released into the UCF as the final lowest-burden remnant. In this picture, the weak force is not a separate force added to the ontology. It is what topological failure in matter looks like.

TECHNICAL NOTE — NEUTRON MASS BALANCE AND BETA-DECAY CASCADE

The intuitive claim is this: the neutron is not a second fundamental baryon. It is a metastable composite whose mass, magnetic moment, and decay products are meant to follow from one geometric picture.

More precisely, UFD treats the neutron as a pressure-locked figure-eight vortex formed when a proton-like source structure binds an oppositely chiral sink-type heavy mode under Genesis conditions. In the present accounting, the combined input is approximately 1876.6 MeV, while the observed neutron mass is about 939.6MeV. The large difference is interpreted as a Coherence Dividend: the energetic payoff of forming the locked composite structure rather than sustaining the two components separately.

Beta decay is then read as a geometric relaxation cascade across the field hierarchy. The composite fails, the proton survives as the lower-strain trefoil endpoint, the electron recoheres in the ULF, and the antineutrino is shed into the UCF as the final lowest-burden remnant. The neutron's magnetic moment is likewise attributed to the composite structure and its opposing chiral organization.

Dr. S wasted no time.

DR. S: All right. We have finally reached the electroweak story. But before we even get to the weak force, I want to know what the neutron is on your view. Because everything you are about to say depends on that.

DR. U: Fair. The neutron is a metastable composite vortex in the UEF.

DR. S: More specifically.

DR. U: At the broadest level, a proton–antiproton hybrid. More carefully, a pressure-locked figure-eight structure formed when a proton-like source mode binds an oppositely chiral sink-type heavy mode under the extreme conditions of the Genesis Event.

DR. S: You are going to have to say that more carefully.

DR. U: The careful version is this. The proton is the low-strain trefoil ground state of heavy matter. Under the immense pressure of the Genesis environment, that structure can bind an oppositely wound heavy mode into one locked composite. The result is not two little particles orbiting one another. It is one topological object whose internal organization has two opposed lobes.

DR. S: So this is where people will immediately hear "proton plus antiproton."

DR. U: And that phrase matters. It should not be erased. But it has to be handled carefully. The claim is not that a free Standard Model antiproton is sitting intact inside the neutron like a bead inside a box. The claim is that the neutron contains an antiproton-like heavy component in the sense of opposite heavy-field chirality, locked into one composite structure.

DR. S: So the proton–antiproton language is real, but not naive.

DR. U: Exactly. Real at the level of source–sink structure. Misleading only if one imagines an ordinary free antimatter baryon trapped whole inside normal matter.

DR. S: So the neutron is composite, but not in the billiard-ball sense.

DR. U: Right. It is one strained topological object.

That answer mattered because the naive picture would have been absurdly easy to dismiss. The claim was not that an ordinary free antiproton sits intact inside the neutron, but that the neutron contains an oppositely chiral heavy component locked into one topological composite.

It did not settle the issue, but it removed the easiest caricature.

Dr. S pressed where he had to next.

The obvious objection.

DR. S: Fine. Then let us get to the first problem. If the neutron is really a proton–antiproton composite in your sense, why is its mass not roughly twice that of a proton?

DR. U: Because neutron formation is a colossal mass-defect event.

DR. S: Meaning?

DR. U: In UFD, when the protonic source structure and the antiproton-like sink structure interlock into the figure-eight neutron, they do not merely add. They reorganize. A tremendous amount of redundant boundary strain is eliminated in the locking process. That released energy is what UFD calls the Coherence Dividend.

DR. S: And how large is this supposed to be?

DR. U: On the order of 936 MeV. That is why the resulting neutron is not sitting up near twice the proton mass. Most of the combined mass-energy is shed in the formation event itself.

DR. S: So the mass objection is not being ignored. It is being made central.

DR. U: It has to be. Otherwise, the composite picture dies immediately.

DR. S: And where does that released energy go?

DR. U: In the original cosmic setting, into the violent radiative output of the Genesis process. That is one reason quasars matter here. In UFD, quasar power is not a side issue. It is one of the natural astrophysical outlets for the enormous coherence energy released when neutrons are forged during the early high-pressure epoch of a galaxy's birth.

That changed the feel of the exchange. The neutron was no longer being treated as a particle that just happens to decay in laboratories. It was being tied to the early forging of matter itself. That made the picture more unified, but also much more exposed.

The neutron had become cosmological.

DR. S: All right. Then tell me about this Genesis setting. Because if the neutron is forged there, I want to know why that matters.

DR. U: It matters because the neutron is not being proposed as an object that forms naturally under ordinary low-pressure conditions. It is a deep-field object. In UFD, there is a Neutron Epoch during the Genesis process, when the environment near the central seed is violent enough to force protonic and antiproton-like heavy vortices into these locked figure-eight structures.

DR. S: So the neutron is born in a forge, not in a calm background.

DR. U: Exactly. A free neutron in ordinary space is like a deep-sea animal brought to the surface. It can exist for a while, but it is no longer in the field environment that made it fully stable in the first place.

DR. S: So Genesis explains not just how the neutron forms, but why it later becomes metastable.

DR. U: Yes. That is one of the central gains of the model.

That sharpened the structure of the whole conversation. The neutron was no longer just a strange alternative picture. It had become the unstable branch of heavy matter, forged in the deepest pressure conditions of cosmic birth and only partially stable afterward.

Only then did Dr. S turn to the weak force itself.

Was the weak interaction a primitive force in its own right, or the visible signature of a deeper instability in matter?

DR. S: All right. Then tell me plainly: what is the weak force?

DR. U: A geometric phase transition in a metastable composite structure.

DR. S: Not a fundamental interaction.

DR. U: Not at the deepest level, no.

DR. S: Then let me state the standard picture as sharply as possible. The weak interaction is not some decorative extra in modern physics. It is one arm of the electroweak theory. It predicts precise branching patterns, parity violation, the W and Z bosons, and a huge amount of collider phenomenology. You do not get to replace all that with "geometric collapse" unless the replacement actually bears weight.

DR. U: I agree.

DR. S: Good.

DR. U: The question is not whether the standard framework is mathematically powerful. It clearly is. The question is whether the weak sector is revealing a separate primitive force, or the instability of a deeper geometry.

DR. S: Naturally you think it is the second.

DR. U: I do.

That was the first real edge of the conversation. Dr. S trusted the standard framework because it had precision and maturity behind it. Dr. U distrusted multiplying primitive sectors when one geometric mechanism might do the same work more intelligibly.

So Dr. S pressed exactly where he had to.

DR. S: Fine. Then why does the neutron decay?

DR. U: Because it is not the ground state of heavy matter.

DR. S: Meaning the proton is.

DR. U: Yes. The proton is the lower-strain trefoil ground state in the dense field. The neutron is a locked composite that can be stabilized in the right nuclear environment, but in free space it carries excess curvature, excess shear strain, and excess topological burden. So when the stabilizing conditions are insufficient, it relaxes.

DR. S: Geometric simplification.

DR. U: Exactly.

DR. S: Into proton, electron, antineutrino.

DR. U: Into trefoil, torus, and genus-zero remnant, yes.

DR. S: Which means beta decay is not fundamentally a quark flavor change mediated by a W boson.

DR. U: Not fundamentally. It is a relaxation cascade across the nested fields.

That answer sharpened the conversation immediately. The weak force was no longer a separate sector. It had become the visible signature of topological failure.

Dr. S heard that clearly and pushed harder.

DR. S: All right. Walk me through the decay itself.

DR. U: The first step is UEF relaxation. The composite figure-eight snaps through into the lower-strain proton configuration. One lobe survives as the trefoil ground state.

DR. S: And the rest?

DR. U: The expelled circulation cannot remain stable in the dense energetic field. It descends into the ULF, where it recoheres into the lowest-strain charged excitation compatible with that field. That is the electron.

DR. S: So the electron is not created from nothing.

DR. U: Exactly. It is recohered circulation shed by the collapsing heavy structure.

DR. S: And the antineutrino?

DR. U: After the electron forms, there is still residual mismatch — twist, writhe, gradient imbalance, however one prefers to say it at this level. That residual content cannot remain stable even in the charged light-matter sector. So it descends one level further into the UCF, where the lowest-burden closed remnant is the genus-zero antineutrino excitation.

DR. S: So the whole process is UEF to ULF to UCF.

DR. U: Yes. That is the key.

For a moment neither of them spoke.

The conversation had now crossed a threshold. This was no longer just a reinterpretation of a force. It was a reinterpretation of one of the most familiar particle transformations in physics as a descent event across the field hierarchy.

Dr. S broke the silence.

DR. S: Let me say what is attractive here. This is not vague. It really is one mechanism. The neutron does not get hit by a separate weak entity. It fails structurally and relaxes into lower-order stable forms.

DR. U: Exactly.

DR. S: That is much stronger than just saying "the weak force is really geometry."

DR. U: It has to be stronger.

DR. S: But the danger is equally obvious. The standard electroweak theory has an enormous amount of precision attached to it. A geometric replacement has to do more than sound cleaner.

DR. U: Of course.

That was the center of the burden. The standard electroweak theory is not just respected for conceptual elegance. It is respected because it organizes a vast body of measured behavior with remarkable numerical precision.

A weaker theory would have been content with conceptual elegance. A stronger one had to face the standard model's precision culture.

So Dr. S moved to the signatures everyone knew.

The W and Z bosons.

The W and Z are not decorative details in the standard model. They are among its most celebrated successes. Electroweak theory predicted them with specific masses before experiment found them. When CERN detected them in 1983 at the expected scale, the discovery was taken as a decisive confirmation of the framework. That is why UFD could not simply deny the W and Z. It had to reinterpret them.

DR. S: Fine. Then what are the W and Z bosons?

DR. U: Not fundamental particles at the deepest level.

DR. S: Naturally.

DR. U: They are resonance signatures of the collapse process.

DR. S: Say that more precisely.

DR. U: When the composite neutron undergoes geometric relaxation, the collapse does not happen silently. The surrounding field rings, the way a struck bell rings. Specific frequencies are excited, and those frequencies depend on the geometry of the structure that is failing. Some of those ringing modes are what particle detectors register as W and Z events. The peaks are real. The signals are real. What changes is what they are signals of. They are not new primitive carriers. They are the field's response to a specific kind of structural failure.

DR. S: So not force carriers.

DR. U: Not primitive carriers, no. More like the fingerprint a particular kind of collapse leaves behind in the field.

DR. S: Which means the resonance peaks are real, but their ontology is different.

DR. U: Exactly.

DR. S: And the W/Z mass split?

DR. U: That is one of the stronger claims in this sector. The W and Z have different masses, and the standard model attributes that to the Higgs structure of electroweak symmetry breaking. UFD treats the split as geometric. A collapsing structure has different characteristic modes depending on its shape and its boundary conditions. The relative spacing comes out naturally from the topology of the failure. The full absolute scale is a harder burden. Getting the ratio is one thing. Getting the exact masses is another, and that part is still developing.

Dr. S leaned back slightly. He did not like the claim. But he could now see what kind of claim it was.

That mattered, because he knew where the standard objection had to go next.

DR. S: All right. Here is the mainstream response. Resonance curves do not by themselves tell you ontology. That much is fair. But electroweak theory does not just say there are resonances. It is a complete mathematical framework. It tells you how the W and Z mix with the photon, how often a given decay produces one outcome rather than another, how parity violation enters at every level, and how all of these features fit together with measurements confirmed to extraordinary precision. So what exactly are you claiming to replace, and what are you still leaving open?

DR. U: Fair question. The core replacement is this. The weak interaction is not primitive. It is the geometric relaxation of a metastable composite. Beta decay is that relaxation, the cascade of reorganization as the strained structure releases excess energy and settles into a lower-strain configuration. The W and Z signatures are the field's response to the moment of collapse. Parity violation is built into the structure itself, not into a separate force law.

The narrator stepped in here because parity violation is one of the most dramatic results in twentieth-century physics, and the general reader needs to feel its weight before Dr. U's reinterpretation can land.

Parity violation is the discovery that weak processes distinguish left from right. Every other known force respects mirror symmetry. The weak interaction does not. Wu's 1956 cobalt-60 experiment showed that beta decay favors one handed orientation over its mirror opposite. In the standard model, that asymmetry is built directly into the weak interaction itself. UFD is trying to relocate that asymmetry. Instead of treating handedness as a primitive property of the force, it derives it from the geometry of the unstable structure that decays.

DR. S: And what remains open?

DR. U: Quite a lot. I am not claiming UFD has reproduced the full quantitative content of the electroweak program. The framework gives a structural picture of the core mechanism, but it has not yet derived the precise numerical values the standard model gets right to many decimal places: exact lifetimes, branching fractions, the detailed W/Z/photon relations, and the rest. What it offers is a different physical picture of where the effects come from. The full precision closure remains a burden it has not discharged.

DR. S: That is a much better answer.

DR. U: It has to be. The core mechanism should be stated strongly. The precision closure should be stated honestly.

That concession improved the conversation. A sloppier theory would have pretended it had already replaced the full electroweak precision program. A stronger one knew the difference between a core mechanism and a complete precision closure.

Dr. S respected that enough to keep going seriously.

DR. S: Let us talk about parity violation directly. The standard theory says the weak interaction has handedness built into it at the deepest level. The mathematical structure treats left and right asymmetrically. What happens to that here?

DR. U: It becomes geometry instead of rule.

DR. S: More specifically.

DR. U: The weak process is asymmetric because the composite neutron is asymmetric. The figure-eight structure has a built-in handedness — a particular twist relative to its mirror image. Once the structure begins to collapse, the available pathways are not symmetric, because the starting point was not symmetric. The asymmetry is not in the force law. It is in the shape of the thing that is coming apart.

DR. S: So parity violation is no longer a surprising property of the force. It is a consequence of the topology of the unstable parent structure.

DR. U: Exactly. A theory that takes the geometry of matter seriously should not need to bolt on parity violation as a separate mystery. The asymmetry in the data is real. But the explanation does not require a primitive force that somehow knows left from right. It requires only that the structure undergoing the decay was not mirror-symmetric to begin with.

For a moment the conversation became quieter.

The dispute had shifted from *what is the weak force?* to *what sort of thing produces asymmetry in the first place?*

That was deeper territory. In the standard model, the weak force is the source of the asymmetry. In UFD, the asymmetry is inherited from the geometry of an unstable composite. That was the right question, even if neither of them yet knew how far it could be carried.

Dr. S then turned to a more concrete concern.

DR. S: All right. If the neutron is composite in your sense, there should be physical consequences besides decay. Magnetic moment, internal dipole structure, maybe environmental effects.

DR. U: Yes.

DR. S: Good. Because that is where a composite story earns its keep.

DR. U: The neutron's magnetic structure should reflect its internal dipole organization. Its lifetime should depend, at least in principle, on environmental agitation and confinement conditions, though I would be careful not to overstate that before the microscopic model is tighter.

DR. S: Again, a better answer than pretending everything is already exact.

DR. U: The composite picture should produce sharper structure, not looser rhetoric.

That was useful. It kept the conversation from floating too high above empiricism. A composite object had to have structure. Structure had to leave traces.

Dr. S pushed on one final pressure point before the conversation could close.

DR. S: Let me tell you what I think is strongest in your picture. You are replacing the weak sector with one geometric postulate rather than a patchwork of separate ingredients. The weak process becomes collapse, parity violation becomes handedness, the electron and neutrino become descent products, and the W/Z signatures become resonance artifacts of the same event. And now, with the Genesis setting included, you are also claiming that the neutron's very formation belongs to the birth of matter itself.

DR. U: Yes.

DR. S: That is a real unification if it works.

DR. U: Exactly.

DR. S: The danger is obvious too. Standard electroweak theory is not just conceptually successful. It is numerically ferocious. So a composite-neutron ontology earns its place only if it eventually does more than simplify the story.

DR. U: Agreed.

DR. S: In other words: elegance is not enough.

DR. U: Never enough.

That was the center of gravity. Dr. U was not merely offering a more intuitive picture of weak decay. He was claiming that one geometric postulate could replace an entire standard sector and connect it to the cosmic birth of matter itself.

Dr. S was not merely defending orthodoxy. He was insisting that replacement-level ambition must eventually survive precision, not just intuition.

Nothing about that was settled.
But the burden had become exact.

DR. S: Let me try the summary test.

DR. U: Go ahead.

DR. S: On your view, the weak interaction is not primitive. It is the geometric relaxation of a metastable composite neutron. The neutron is a figure-eight-like UEF structure with built-in chirality and strain — at the broadest level a proton–antiproton hybrid, more carefully a pressure-locked source–sink composite rather than a free antiproton trapped inside ordinary matter. It is forged in the extreme conditions of the Genesis Event, and its formation releases an enormous Coherence Dividend that helps answer the mass-disparity problem and may connect to quasar power. In beta decay, the proton survives as the lower-strain trefoil ground state, the electron recoheres in the ULF as the shed toroidal circulation remnant, and the antineutrino descends into the UCF as the final simplest coordination remnant. The W and Z signatures are not fundamental carriers but resonance modes of the collapse, and parity violation reflects the handedness of the unstable structure rather than a separately postulated asymmetry in the force.

DR. U: That is fair.

DR. S: The explanatory gain is genuine.

DR. U: Yes.

DR. S: Genuine gains create genuine obligations.

DR. U: Yes.

DR. S: Good. Then neither of us can pretend this is easy.

INTERLUDE

By the end of this exchange, the weak sector no longer looked like an isolated interaction.

At least in the UFD picture, it had become the place where topological descent turns into transformation physics.

The neutron mattered because it was not a second equally stable heavy-matter endpoint. It was a strained composite, formed when heavy-field geometry locked into a higher-burden arrangement than the proton could sustain in isolation. The broader context made that even more striking: this was not only a story about decay in the present universe, but about the forging of matter during the Neutron Epoch of the Genesis process, when immense coherence energy was released as the first neutrons formed.

That is a strong unifying move.

It is also exposed in exactly the right place: if the composite-neutron picture does not eventually organize parity, decay, resonance structure, environmental signatures, and the energy bookkeeping of neutron formation in a disciplined way, then the elegance of the reinterpretation will not be enough.

141

Dr. U succeeded in making the weak sector physically legible. The neutron became a genuine structural object, weak decay became geometric collapse, parity violation became handedness, the daughter products became the natural topological descendants of a strained composite, and the mass-disparity objection was answered by the Coherence Dividend released in Genesis-era neutron formation.

Dr. S succeeded in holding the proper line of pressure: a unified geometric story is not enough. The replacement earns its place only if its elegance tightens into real structural and, eventually, precision constraint.

OF COURSE, IF THE WEAK FORCE IS REALLY A TOPOLOGICAL TRANSITION AND PARTICLE IDENTITY IS GEOMETRIC, THEN THE NEXT QUESTION IS WHETHER ELECTROMAGNETISM ITSELF CAN BE MADE EQUALLY PHYSICAL.

CHAPTER 11

Hydro Electrodynamics

Charge, magnetism, and force as flow

Electromagnetism is one of the most successful theories in the history of science. It is also one of the most familiar. Every time you turn on a light, send a text message, listen to a radio, or pull open a refrigerator door with a magnet attached, you are seeing the same underlying interaction at work. Light, radio waves, electric current, and magnetism all belong to one framework. That unity is one of the great achievements of physics.

The story of how that unity was discovered matters because it sets the stage for what UFD is trying to reinterpret. In the early nineteenth century, electricity and magnetism were treated as separate phenomena. Electricity concerned charges and currents. Magnetism concerned lodestones, compass needles, and iron. Then James Clerk Maxwell wrote down the equations that unified them. Those equations did more than describe electricity and magnetism together. They predicted that disturbances in the combined field should propagate as waves, and that the speed of those waves should equal the measured speed of light. Light, it turned out, was electromagnetic. That remains one of the deepest unifications in all of science.

In the twentieth century, the framework was pushed still further. Quantum electrodynamics re-described electromagnetic interaction in terms of photons and a coupling strength set by the fine-structure constant. The resulting theory has been confirmed to extraordinary precision. In some domains, it is accurate to more decimal places than almost anything else human beings have ever measured. The mathematics is breathtaking. The empirical success is not in doubt.

And yet some of the deepest physical questions remain strangely untouched. What is charge, physically? Why are there exactly two signs of charge? Why do opposites attract and likes repel? Why is magnetism so tightly connected to electricity that the two are really aspects of one field? Why should the force fall off with distance in just the way it does? In the standard picture, the usual answer is ultimately formal: charge is what enters the equations as a parameter, the field behaves as Maxwell's equations say it behaves, and the photon is the object the formalism uses to describe interaction. The theory tells you what electromagnetism does with astonishing success. It does not always tell you what electromagnetism is.

That is why UFD's ambition here is risky. It is not trying to repair a broken theory. It is trying to make a successful theory more physically intelligible. It wants to keep the successful equations while changing the ontology underneath them.

On the UFD view, charge is not a primitive property that particles simply possess. It is a pattern of circulation in a real field. A positive charge is a source-type vortex: a structure in which the surrounding field flows outward in a particular handed pattern. A negative charge is a sink-type vortex: the corresponding inward form. The two kinds of charge are not arbitrary labels. They are the two possible chiralities of a vortex in the field, much as a screw can be threaded left-handed or right-handed and no third option exists.

Magnetism is then no longer a separate mystery that somehow appears when charges move. It is the organized rotational flow those moving vortices generate in the surrounding medium. When charges move coherently as a current, the field around them organizes into a circulating pattern. That circulating pattern is what we measure as a magnetic field. Electricity and magnetism are not two phenomena later stitched together by clever mathematics. They are two aspects of one underlying field behavior.

The electromagnetic force itself becomes a pressure-based effect of shared flow in the medium. Opposite charges attract because the field between source and sink organizes into a pressure differential that pulls the structures together. Like charges repel because the surrounding flow organizes differently and drives them apart. On this view, the Coulomb force is not action at a distance imposed by axiom. It is the hydrodynamic consequence of source and sink structures sharing one medium.

And Maxwell's equations? UFD does not seek to discard them. It seeks to derive them. The claim is that they are the small-amplitude hydrodynamics of the communicative field itself — the equations governing how a real continuous medium responds to circulation, sources, and sinks. What Maxwell discovered in the nineteenth century, on this view, is not a set of irreducible commandments. It is the ordinary field behavior of a real medium at the scales where the medium behaves linearly.

Dr. S began where he always did when the subject threatened to become too intuitive too quickly.

DR. S: All right. Let us see if I understand the ambition here. You are not just reinterpreting electromagnetism. You are trying to replace QED's basic ontology with hydrodynamics in the ULF.

DR. U: With hydrodynamics, yes. Not as a loose analogy, but as the physical basis.

DR. S: Then let me state the standard objection immediately. QED is not just successful. It is one of the most precise theories ever constructed. You do not get to say "charge is really flow" and claim progress because it sounds more tangible.

DR. U: I agree.

DR. S: Good.

DR. U: The question is not whether QED works. It clearly does. The question is whether its basic ingredients — charge, magnetism, coupling, even the electromagnetic force itself — are primitive, or whether they are the visible behavior of a deeper field.

DR. S: Naturally you think it is the second.

DR. U: I do.

That was the first real point of strain.

Dr. S trusted the formal strength of electromagnetism enough not to demand more from it.

Dr. U trusted the formal strength too, but took that very success as a reason to ask what the formalism was actually describing.

So Dr. S went to the first hard question.

Charge.

This mattered because charge is usually introduced as a basic label: positive, negative, or neutral. UFD was trying to replace that label with a geometric property of the field itself.

DR. S: Fine. What is electric charge?

DR. U: Chirality.

DR. S: That is too compressed. Say it properly.

DR. U: A stable vortex in the ULF is not neutral with respect to the medium if it has a definite handedness. One chirality behaves as a source: it pushes the surrounding ULF outward. The opposite chirality behaves as a sink: it draws the ULF inward. Those are what we call positive and negative charge.

DR. S: So charge is not a mysterious intrinsic label attached to a particle.

DR. U: No. It is the field-level expression of how that vortex is wound.

DR. S: Source and sink.

DR. U: Exactly.

DR. S: And charge quantization?

DR. U: It follows from quantized circulation. The ULF does not permit arbitrary winding. Stable circulation comes in discrete topological units, so charge does too.

That answer changed the feel of the conversation immediately.
Charge was no longer being treated as a label written onto matter from nowhere.
It had become a geometric property of a real vortex in a real field.
That did not make it true.
But it made it legible.

Dr. S pushed at once to the next obvious question.

Force.

DR. S: All right. Then what becomes of Coulomb's law? Because charge may be more vivid now, but that does not yet tell me why opposites attract and likes repel.

DR. U: In UFD, the Coulomb interaction is preserved mathematically while reinterpreted physically as field stress in the ULF.

DR. S: Meaning?

DR. U: Like-charged vortices impose incompatible deformations on the surrounding medium, so the field between them remains stressed and they repel. Oppositely charged vortices permit partial field relaxation. The ULF can flow from source to sink, reducing strain and creating an effective low-pressure channel between them. That is attraction.

DR. S: So the force becomes a field-pressure effect.

DR. U: Exactly. The mathematics stays Coulomb. The mechanism changes.

DR. S: That is one of your stronger moves.

DR. U: It should be. A theory that makes charge physical should not leave force mysterious.

That shifted the ground.

A weaker reinterpretation would have made charge more vivid but left the force as primitive as before.

A stronger reinterpretation made the force a downstream consequence of the same chirality claim.

Dr. S moved immediately to magnetism.

Magnetism is often introduced to students as something distinct from electricity and only later unified with it mathematically. UFD was trying to make that unity physical from the start: source-and-sink behavior on one side, rotational flow on the other, both in one field.

DR. S: Fine. Then what is magnetism?

DR. U: Vorticity in the ULF.

DR. S: Again, say it properly.

DR. U: A moving or spinning chiral vortex drags the surrounding ULF into coherent rotational flow. That induced rotation is what we call the magnetic field.

DR. S: So magnetic field lines become—

DR. U: Streamlines of ULF circulation.

DR. S: Not abstract lines we draw for convenience.

DR. U: They are still useful drawings, but yes — physically, they represent the rotational structure of the field itself.

DR. S: Which means magnetism is not a second primitive force layered on top of electricity.

DR. U: Exactly. It is the vorticity response of the same medium.

DR. S: So electrostatics is source-and-sink behavior, magnetism is rotational behavior, and both belong to one field.

DR. U: Yes.

That was the first place where the architecture of the conversation became fully visible.

Charge and magnetism were no longer separate mysteries. They were different kinematic faces of one field.

Dr. S saw that immediately and pushed to the most familiar equation-level burden.

Maxwell's equations.

Maxwell's equations are not just another set of formulas. They are the classical backbone of electromagnetism and one of the great unifications in physics. If UFD wanted to change the ontology underneath electromagnetism but keep the equations, it had to show that the equations emerged from the medium rather than surviving by verbal permission.

DR. S: All right. Then what happens to Maxwell? Because if you tell me the equations survive but the ontology changes, I want to know how deep the survival goes.

DR. U: The survival goes very deep. In the weak-field regime, Maxwell's equations emerge as the linearized field equations of the ULF.

DR. S: So not postulated from nowhere.

DR. U: No. Recovered as the low-amplitude hydrodynamics of the communicative field.

DR. S: Then take them one at a time. Gauss's law.

DR. U: Source chirality creating divergence in the effective coarse-grained field.

DR. S: Say that less compactly.

DR. U: A positive charge is a source-type vortex. A negative charge is a sink-type vortex. If you surround such a structure with an imaginary surface and ask how much field crosses that surface, you recover exactly the logic of Gauss's law. Divergence is just the mathematical way of saying that a source or sink sits inside the surface.

DR. S: Faraday?

DR. U: A changing rotational pattern in the field induces a linear response around it. That is Faraday's law: changing magnetic structure communicates itself outward as induced electric circulation.

DR. S: Ampère-Maxwell?

DR. U: The same effect in the other direction. Drag a charged vortex through the field, or change the electric field rapidly enough, and the surrounding medium develops rotational flow. That is the magnetic response.

DR. S: And no magnetic monopoles?

The absence of magnetic monopoles is one of the standing oddities of electromagnetism. Electric charges exist as isolated positives and negatives. Magnetic poles do not. The standard framework tolerates that as an empirical fact. UFD tries to make it structural.

DR. U: Exactly. In UFD, the absence is structural. The magnetic sector is vorticity — circulation in the field. Vorticity does not have sources and sinks the way charge does. A whirlpool does not have a positive end and a negative end. It has only its rotation, and that rotation closes on itself by definition. There cannot be a magnetic monopole because there cannot be a half-whirlpool.

Dr. S nodded once.
Not agreement.
Recognition.

The exchange had done something rare. It had taken four equations that mainstream physics treats as unified by formal beauty alone and given each of them a physical reason to exist. Maxwell had survived intact, but he had stopped being a list of axioms. He had become a description of what the field does.

This was now a real replacement claim, not just a metaphor.
So he pushed exactly where such claims usually become most helpful for ordinary intuition.

He asked for a picture.

DR. S: Fine. Give me the physical picture. Not the equations. The picture.

This was the right demand. A theory that claims deeper physical intelligibility should be able to say more than "the equations still work." It should be able to picture what the equations are equations of.

DR. U: Think of a whirlpool in water.

DR. S: Go on.

DR. U: A magnetic field is like a stable rotational eddy in the ULF. If that eddy speeds up, slows down, or shifts, it disturbs the surrounding medium and creates a linear flow. That is the physical picture behind Faraday's law.

DR. S: And the reverse?

DR. U: A changing electric field is an accelerating linear flow in the ULF. When you drive fluid hard in one direction, you naturally generate rotational structure around it. That is the physical picture behind the Ampère-Maxwell law.

DR. S: So the electric and magnetic fields keep generating one another because linear and rotational flow keep bootstrapping one another in the same medium.

DR. U: Exactly.

DR. S: Which means light itself is just that self-sustaining cycle.

DR. U: Yes. A photon is the leapfrogging of linear and rotational excitation in the ULF.

That answer did more than clarify Maxwell.
It linked electromagnetism back to light without changing the subject.
The electric and magnetic fields were no longer two separate mathematical actors that happened to dance well together.
They had become two aspects of one moving medium.

Dr. S then turned to one of the most concrete payoffs in the whole theory.

The neutron.

DR. S: All right. Then explain the neutron's magnetic moment.

DR. U: One of the best examples in the framework.

DR. S: Good. Because standard physics has to work hard there. The neutron has no net charge, yet it has a substantial magnetic moment.

DR. U: In UFD, that becomes physically straightforward. The neutron is a composite proton–antiproton source–sink dipole. Its net electric charge cancels, but its internal dipole structure does not. When that internal source–sink structure spins, it drives a coherent rotational eddy in the surrounding ULF. That eddy is the neutron's magnetic moment.

DR. S: So the neutron is electrically neutral because the source and sink balance, but magnetically active because the dipole still spins.

DR. U: Exactly.

DR. S: That is a very clean payoff.

DR. U: It should be. A composite ontology should explain why neutrality and magnetism can coexist.

This was one of the chapter's clearest structural payoffs. The neutron's neutrality and magnetism no longer had to be tolerated together as a formal fact. They followed naturally from an internal source–sink dipole that spins.

Dr. S respected that.
Which was why he now pushed on the standard ontology itself.

DR. S: Fine. Then what becomes of virtual photons?

DR. U: The formalism remains useful. The ontology changes.

DR. S: Meaning?

DR. U: In the standard framework, virtual photons are the bookkeeping language of interaction. In UFD, the underlying process is not exchange between point particles in emptiness, but transient pressure waves and coherent coupling in the ULF. The formal language survives because it tracks the same interaction structure. The physical picture underneath it is field dynamics rather than abstract exchange.

DR. S: So you are not discarding the calculation.

DR. U: No. I am changing what the calculation is about.

That answer mattered because it kept the conversation disciplined.
The best UFD frameworks never tried to erase successful mathematics.
They tried to say what the mathematics was a mathematics of.

Dr. S then moved to the last big burden.

Precision.

DR. S: Let me tell you what is strongest here. You are not merely saying electromagnetism "feels like a fluid." You are trying to derive its formal unity from one field picture: charge from chirality, magnetism from vorticity, force from pressure and circulation, Maxwell from linearized hydrodynamics.

DR. U: Yes.

DR. S: That is a genuine unification if it works.

DR. U: Exactly.

DR. S: The danger is also obvious. If the field picture is real, then it must eventually constrain the places where the point-particle abstraction becomes suspect. Otherwise, it is just an after-the-fact story.

DR. U: Agreed.

DR. S: So where would it show up?

DR. U: In the places where the standard ontology should begin to fray. Finite-core deviations from perfect point-source behavior at ultra-short scales. No elementary magnetic monopoles. Electromagnetic waves as genuine ULF waves. Possibly tiny propagation anisotropies if the large-scale ULF is not perfectly uniform.

DR. S: So the reinterpretation has teeth.

DR. U: It has to.

For a moment neither of them spoke.
The issue was no longer whether electromagnetism could be made more intuitive.
It had become whether one of the most successful theories in science was already the visible hydrodynamics of a deeper field.

Dr. S still trusted the standard theory more than the replacement.
Dr. U still distrusted any theory that treated charge and force as brute.
Nothing about that was settled.
But the target was now exact.

DR. S: Let me try the summary test.

DR. U: Go ahead.

DR. S: On your view, electric charge is the field-level manifestation of vortex chirality: source-type circulation gives positive charge and sink-type circulation gives negative charge. Magnetism is coherent rotational flow — ULF vorticity — generated by moving or spinning chiral vortices. The Coulomb interaction is the pressure-based effect of compatible or incompatible field deformation. The Lorentz force is the combined pressure-gradient and vorticity response of the medium. Maxwell's equations are not primitive axioms but the weak-field hydrodynamics of the ULF. There are no magnetic monopoles because vorticity has circulation, not source topology. And even the neutron's magnetic moment follows naturally from its internal proton–antiproton dipole structure.

DR. U: That is fair.

DR. S: The attraction is obvious.

DR. U: Yes.

DR. S: The danger is equally obvious.

DR. U: Yes.

DR. S: Good. Then the burden is finally clear.

INTERLUDE

By the end of the exchange, electromagnetism no longer looked like a list of primitive facts.

At least in the UFD picture, charge, magnetism, force, and Maxwellian propagation had all been drawn into one field ontology.

That was the gain.

Charge stopped being a bare label and became chirality. Magnetism stopped being a separate mystery and became rotational flow. Coulomb attraction and repulsion stopped being unexplained rules and became the pressure behavior of a real medium. The mutual generation of electric and magnetic fields stopped being a formal symmetry without a picture and became the interplay of linear and rotational flow. Even the neutron's magnetic moment, one of the most familiar puzzles in particle physics, became a direct consequence of internal source–sink structure.

The burden was equally strong. A vivid ontology is not enough. If electromagnetism is really the hydrodynamics of light, then the framework must preserve the exact success of Maxwell and QED at accessible scales while exposing itself honestly where a finite-core, medium-based ontology should begin to diverge.

That was what made the chapter dangerous in the right way. It did not merely offer a nicer story about familiar equations. It claimed that some of the most successful equations in physics might already be the visible hydrodynamics of a deeper light-bearing medium.

TEMPORARY VERDICT

Dr. U succeeded in making electromagnetism physically legible. Charge became chirality, magnetism became vorticity, force became flow,

Maxwell's equations became the weak-field hydrodynamics of a real medium, and the neutron's magnetic moment became a clear structural payoff of the composite picture.

Dr. S succeeded in holding the decisive line of pressure: a vivid ontology is not enough. It earns its place only if it preserves the unmatched formal success of electromagnetism while tightening the physics where the standard ontology still looks primitive.

If light, relativity, weak decay, and electromagnetism can all be drawn into one field-based picture, then the next burden is the quantum formalism itself: what is the wavefunction, what is measurement, and what is entanglement when the field is real?

CHAPTER 12

The Wavefunction, Measurement, and Entanglement

Probability tool or real field structure?

Modern quantum mechanics is the most successful theory in science, but it has always been difficult to picture physically. Its equations predict interference, tunneling, atomic structure, chemical bonding, and measurement statistics with astonishing precision. What they do not settle, at least in the standard reading, is what the wavefunction is in the world. Is it only a calculation device, or does it correspond to something physically real?

UFD answers that question directly. It treats the wavefunction as a real pattern of amplitude and phase in the Universal Light Field. On this view, particles are not bare points with probabilities attached to them. They are stable localized structures in that field. Superposition means that a real medium can support more than one resonant mode at once. Measurement is then no longer mysterious bookkeeping, but a physical localization event involving decoherence, irreversible coupling, and mode selection.

That is a very large claim. Quantum mechanics is not a theory one reinterprets casually. The standard formalism works extraordinarily well, and the history of quantum foundations is crowded with interpretations that sound satisfying until they are forced to say exactly what in the mathematics corresponds to what in the world. Any realist reading has to do more than provide a vivid picture. It has to show that the picture makes the formalism more intelligible without quietly discarding what made the formalism successful in the first place.

UFD's wager is that the gain here is not merely philosophical. If the wavefunction is real field structure, then several familiar quantum puzzles begin to change character. Wave-particle duality becomes one field organization showing two faces. Collapse becomes a real physical event rather than a primitive interruption in the rules. Entanglement becomes a non-separable field structure rather than a mysterious link between independent little objects. And if that ontology is right, then coherence itself may become a more direct target of control rather than only of statistical management.

Maturity note. This chapter is best read as a more physical picture of quantum theory, not as a finished replacement for the full quantum formalism. But it is more than interpretation alone. In the nonrelativistic regime, UFD treats the Schrödinger equation as the effective evolution law of a real oscillatory mode in the ULF. That does not rebuild all of quantum mechanics from first principles, but it does give the ontology firmer grounding than standard instrumental language usually provides. More ambitious extensions, especially those involving the UCF in irreversible localization, remain secondary and less mature than the core ULF picture.

TECHNICAL NOTE — THE WAVEFUNCTION AS REAL FIELD STRUCTURE

The intuitive claim is simple: the wavefunction is not just a line in a calculation. It is the shape of a real disturbance in a physical medium. On this view, quantum theory works because it tracks what the field is actually doing.

More precisely, the Resonant Field Interpretation treats the wavefunction as a real pattern of amplitude and phase in the Universal Light Field, not merely as a bookkeeping device. In the nonrelativistic regime, the wavefunction can be understood as the slowly varying shape of a real oscillatory mode in that field, and the Schrödinger equation describes how that shape evolves over time. This is not yet a full rebuilding of quantum theory from first principles. But it is more than a change of language. It is a claim about what the formalism is a formalism of.

Dr. S had been waiting for this one.

Quantum mechanics had a way of embarrassing everyone equally.
The mathematician, because the equations worked too well to ignore.
The realist, because the ontology often seemed to evaporate when pressed.
The philosopher, because every interpretation sounded cleanest until it had to face the actual physics.

Dr. S did not mind mystery when it was earned.
What he distrusted was the habit of treating ontological vagueness as though it were a sign of sophistication.

So he began where he had to begin.

DR. S: All right. We have delayed the hard part long enough. What is the wavefunction?

DR. U: A real pattern in a field.

This was the conversation's first unavoidable pressure point. If the wavefunction is only a calculation device, then quantum theory may not owe us a deeper picture. If it is physically real, then almost everything else changes with it.

DR. S: Not a probability tool.

DR. U: Not fundamentally, no.

DR. S: Then let me state the standard caution before you get too comfortable. Quantum mechanics works because it gives us a disciplined formalism. The trouble begins the moment people decide they know what the formalism really means. That is where a century of interpretive overreach begins.

DR. U: I know the history.

DR. S: Then you know why I am cautious. Quantum ontology has a bad habit of sounding profound while adding confusion.

DR. U: Sometimes. But refusing ontology altogether is not the same as discipline. Very often it just freezes the question in place.

DR. S: Fair enough. Then give me the strong version. When I write down a wavefunction, what am I writing down on your view?

DR. U: The amplitude and phase pattern of a real excitation in the ULF.

DR. S: So the wavefunction is not information about a system. It is the system.

DR. U: In the relevant sense, yes.

DR. S: And the particle?

DR. U: A localized coherent excitation supported by that field.

DR. S: A vortex again.

DR. U: Often, yes. Or more generally a stable topological mode. The key point is that the particle is not a point with a wave attached to it. It is one field-organization seen in two ways at once: extended in its mode structure, localized in its coherent core.

DR. S: So wave-particle duality becomes one structure showing two faces.

DR. U: Exactly.

Dr. S did not object immediately. He objected carefully, which was usually worse.

DR. S: Let me tell you what worries me. Your picture sounds satisfying because it replaces abstraction with image. The wavefunction becomes a real field. The particle becomes a coherent excitation. Collapse becomes disruption of resonance. Fine. But replacing austere language with vivid language is not yet explanatory progress.

DR. U: I agree.

DR. S: Good.

DR. U: The point is not to make quantum mechanics friendlier. The point is to ask what sort of world would make the formalism natural. If interference is real, what is interfering? If bound states are stable, what is standing? If localization occurs, what is being localized?

DR. S: And your answer in every case is: a field.

DR. U: Yes.

DR. S: That is at least clean.

DR. U: It should be. A confused ontology is worse than no ontology.

For a moment the conversation quieted.
The disagreement was no longer about whether quantum mechanics worked.
That was beyond dispute.
It was about whether success entitled one to stop asking what the success was about.

Dr. S pushed directly into the hardest point.

Measurement.

The measurement problem is the deepest and most embarrassing puzzle in standard quantum mechanics. The wavefunction evolves smoothly according to Schrödinger's equation, spreading into superpositions of multiple possible outcomes. But experiments do not show all the outcomes at once. They show one definite result. The usual formalism therefore adds a second rule — collapse — without fully explaining what triggers it or what physically happens when it occurs.

DR. S: Fine. Then what is measurement?

DR. U: A physical localization event.

DR. S: More specifically.

DR. U: A wavefunction in UFD is a real field configuration — a pattern of amplitude and phase spread across the region the system occupies. When the field is left alone, it can evolve into a state containing multiple live possibilities at once, the way a vibrating string can carry several harmonics simultaneously. Measurement is an interaction strong enough to disrupt that combined state. When the field couples irreversibly to a detector or to a larger environment, the coupling forces the field to settle into one of the configurations the apparatus can register.

DR. S: So collapse is real.

DR. U: As a physical event, yes.

DR. S: Not as a primitive axiom.

DR. U: No. The wavefunction does not collapse because someone looked at it. It collapses because the field has been forced into a specific configuration by interaction with another physical system.

DR. S: And no observer is required.

DR. U: None. Measurement is not consciousness staring at a system. It is coherence breaking under physical coupling. A detector, a photographic plate, a Geiger counter — any of these can force localization. The observer reads the result afterward.

Dr. S leaned forward slightly. He had been waiting to deliver the next point.

DR. S: Here is the standard objection. Decoherence already gives a respectable account of why interference disappears in practice. What it does not give you is why one definite outcome occurs rather than another. So when you say "physical localization," I want to know whether you have actually solved the hard part or just renamed it.

Decoherence is one of the most important advances in quantum foundations. It explains why interacting with a complex environment destroys observable interference between different branches of the wavefunction. It explains why large objects look classical. But it does not by itself explain why one branch becomes the one actually realized outcome.

DR. U: Fair question. Decoherence is real, and UFD does not deny any of it. What decoherence tells you is why interference between branches becomes unrecoverable. What it does not tell you is why one specific branch is realized rather than another. UFD's claim is that if the field is real, then localization is also a real physical event. The full state of the field, together with the geometry of the interaction and the state of the environment, determines which branch the field actually settles into.

DR. S: Which sounds deterministic.

DR. U: Not in the simple sense. I am not claiming that quantum probabilities disappear or that the entire microscopic account of selection has been worked out. The Born rule still gives the right statistics. What changes is the picture of what produces those statistics. In the standard view, the probabilities are fundamental. In UFD, they reflect how a real field with real configurations and real interactions distributes into definite states under measurement, even if the full microscopic mechanism of individual selection is not yet completely worked out.

DR. S: Better.

DR. U: It has to be said that way. The stronger claim is that collapse is not primitive and not observer-triggered. Selection is a physical process, not a magical jump. The deeper question of how each individual outcome is selected by the full state of the field is one I do not pretend to have completely answered. What I am saying is that the question now has a place to live.

That mattered.

A less honest presentation would have pretended to solve the measurement problem outright.

A stronger version knew the difference between making collapse physical and making every last detail of outcome selection mathematically final.

Dr. S respected that enough to keep pressing seriously.

DR. S: All right. Then what about entanglement? Because this is where quantum theory most obviously resists ordinary physical pictures.

Quantum entanglement is the phenomenon in which two systems that have once interacted remain correlated even when separated by vast distances. Standard quantum mechanics predicts those correlations precisely, but the physical picture has always been controversial.

DR. U: In UFD, entanglement is not primarily a mysterious connection between independent particles. It is a single non-separable field configuration.

DR. S: Meaning?

DR. U: Two systems that have interacted can no longer be treated as two fully separate field structures. They become one extended configuration of the medium whose state does not factor cleanly into independent parts. There are no longer two particles with two separate field patterns. There is one combined pattern spread across two regions of space.

DR. S: So the nonlocality is structural, not signaling.

DR. U: Exactly. The field can be extended and non-separable without any need for faster-than-light messages passing back and forth. Nothing is being sent from one side to the other when a measurement is made. There is only one underlying configuration, and the measurement reveals correlated features of that single configuration.

DR. S: That is one of your cleaner moves.

DR. U: It should be. Apparent nonlocality becomes much less paradoxical once one stops imagining two little separate objects somehow whispering to one another across empty space.

This was one of the cleanest gains of the chapter. Entanglement became less mysterious once the theory stopped imagining separate objects that must somehow remain secretly connected and instead treated the system as one non-separable field structure from the start.

For a moment the conversation became clearer rather than stranger.

Entanglement had stopped sounding like magic and started sounding like what UFD wanted it to sound like: one field that had not really split into independent pieces.

Dr. S then pressed on the obvious mathematical worry.

Configuration space.

When quantum mechanics describes many-particle systems, the formalism is written in a high-dimensional mathematical space rather than directly in ordinary three-dimensional space. That creates one of the deepest philosophical puzzles in quantum theory: if the wavefunction is real, does reality itself live in this bizarre high-dimensional arena?

DR. S: Fine. But standard quantum mechanics is written in configuration space. You are talking as though the real story takes place in ordinary three-dimensional field space. Are you just ignoring the mathematical problem?

DR. U: No. I am reclassifying it.

DR. S: Go on.

DR. U: Configuration space remains useful mathematics. It is the right tool for calculating how many-particle systems behave, and UFD does not propose to abandon it. But in UFD, configuration space is representational rather than fundamental. It is a way of encoding joint constraints on a single underlying field that lives in ordinary three-dimensional space.

DR. S: So the high-dimensional formalism stays, but the ontology underneath it changes.

DR. U: Exactly. The equations are not being discarded. What changes is what the equations are taken to describe. Standard quantum mechanics treats the configuration-space wavefunction as fundamental. UFD treats it as a mathematical encoding of a more basic three-dimensional field configuration.

That answer kept the discussion on the right track.
The point was not to overthrow the quantum formalism.
It was to stop treating its most abstract elements as automatically ontological.

Dr. S then moved to the one part of the framework that had to be handled with particular care.

The Universal Coordination Field.

The UCF is the most speculative element of the UFD framework. It is not where the core quantum reinterpretation should stand or fall.

DR. S: All right. Now let us get to the dangerous part. You have a real ULF ontology for the wavefunction, collapse, and entanglement. Fine. Where, if anywhere, does the UCF enter?

DR. U: Carefully.

DR. S: Good start.

DR. U: The core quantum picture does not need the UCF in order to say that the wavefunction is real, that particles are field excitations, that collapse is physical localization, or that entanglement is non-separability. All of those claims belong to the ULF story. The reinterpretation this chapter is developing stands on the ULF picture alone.

DR. S: And the stronger claim?

DR. U: The stronger claim is only conditional. The UCF may provide a deeper coordination layer during irreversible localization — something that helps enforce global phase consistency without acting as a signaling channel. If that turns out to be true, it would give UFD a more unified picture of measurement than the ULF story alone provides. But it is not required.

DR. S: So it is supplementary, not foundational, at this stage.

DR. U: Exactly. The core ontology should stand on the ULF picture. The UCF extension is more ambitious and less mature, and it has to be presented that way.

DR. S: That distinction matters a lot.

DR. U: I know. A theory that mixed its strongest claims with its most speculative ones would deserve to be dismissed.

That improved the whole conversation.
The chapter became stronger by keeping the core quantum ontology in the ULF and treating the UCF only as a more ambitious, less mature extension.

Dr. S then widened the frame.

DR. S: Let me be fair. There is something genuinely attractive here. You are trying to preserve the mathematical success of quantum mechanics while removing several of its deepest ontological frustrations. The wavefunction stops being just a tool for predicting probabilities and becomes a real configuration of a real medium. Collapse stops being a postulate added by hand and becomes a physical event. Entanglement stops being spooky action between separate little things and becomes a single field structure spread across multiple regions. Each of these moves removes a source of genuine philosophical discomfort that working physicists have been carrying for almost a century.

DR. U: Yes.

DR. S: And the danger?

DR. U: That the new picture could sound more satisfying without actually clarifying enough. A reinterpretation that feels good but does not generate any new content is just a different story told over the same equations. It would be a philosophical preference, not a scientific advance.

DR. S: Exactly.

DR. U: Which is why the view has to be judged carefully. At present, this is mainly an interpretive and explanatory gain. It is not a claim that every quantum number or every measurement statistic has been rebuilt from scratch. The standard formalism still does most of the calculating work. What changes is the picture of what the formalism describes.

DR. S: Better.

DR. U: It has to be said that way. Otherwise, the ontology would overpromise.

DR. S: Fine. But so far this still sounds mainly interpretive. What have you gained besides a cleaner picture?

DR. U: Potential leverage.

DR. S: Meaning?

DR. U: If the wavefunction is a real field structure, and if decoherence is a real breakdown of resonance in that field rather than an irreducible black box, then quantum control changes. Noise is no longer just something to average over statistically. It becomes something that can be mapped, countered, and, in principle, actively stabilized.

DR. S: You mean technologically.

DR. U: Yes. If coherent quantum states are real resonant configurations of the field, then one should be able to look for ways of reinforcing them directly — methods that actively support the coherence of a quantum system rather than merely shielding it from disturbance. That is the basic idea behind what UFD calls Resonant Stabilization.

DR. S: So you are saying the interpretation is not only philosophical. It points toward a different engineering program.

DR. U: Exactly.

DR. S: And you think that matters as evidence?

DR. U: It matters as exposure. If the ontology is right, it should generate practical consequences. If carefully designed attempts at Resonant Stabilization fail across the board, then this view loses one of its strongest claims to physical relevance. If, on the other hand, they succeed in ways that standard control language alone does not naturally suggest, then the interpretation stops being merely interpretive.

This was where the ontology stopped being only philosophical. If quantum coherence is a real resonant-field phenomenon, then the interpretation should begin to change which control strategies look physically promising.

A weaker theory would have been content to make quantum mechanics feel less embarrassing.
A stronger one had to show that a new ontology might alter not only how the theory is understood, but what becomes buildable.

Dr. S recognized that immediately.

Which was why he sharpened the point instead of dismissing it.

DR. S: Let me put the pressure where it belongs. Quantum engineering already exists. Coherence protection, error correction, pulse shaping, dynamical decoupling — none of that requires a new ontology. So why should I treat your Resonant Stabilization idea as more than a poetic gloss on work people are already doing?

DR. U: Because the ontology tells you what the engineering is engineering. In the standard picture, many of those techniques are operational strategies layered over a formalism. In this picture, they become targeted interventions in a real resonant medium. That changes how one thinks about what is possible, what should be measured, and what kinds of control might matter most.

DR. S: Such as?

DR. U: Such as coherence floors, topology-sensitive decoherence rates, field-geometry effects that ordinary qubit language may hide, or active resonant reinforcement that stabilizes a state not merely by isolating it, but by supporting the mode itself.

DR. S: So again the claim is not that the mathematics of quantum information is discarded.

DR. U: No. The claim is that the physical picture underneath it becomes more specific, and therefore more experimentally vulnerable.

That improved the whole exchange.
The issue was no longer only whether the quantum discussion made the formalism easier to picture.
It had become whether the ontology, if taken seriously, pointed toward testable consequences outside pure interpretation.

That did not prove the ontology.
But it made the stakes more scientific.

DR. S: Let me ask the last question the way a skeptic should. What, in the end, has actually been gained here that standard quantum mechanics does not already give me?

DR. U: A physical world.

DR. S: That is a good sentence. It is also dangerous. Explain it.

DR. U: For nearly a century, quantum mechanics has been the most successful theory in the history of science and the most evasive about what it describes. Generations of physicists have been told some version of "shut up and calculate" — that asking what the wavefunction is, or what happens during measurement, or whether entanglement involves anything physical, is either a category mistake or a waste of time. The formalism predicts everything. The world it describes has been left systematically underspecified.

UFD tries to stop that drift. It says the world is made of real fields. The wavefunction is one of their structures. Collapse is something that happens to them. Entanglement is what it means for them not to have split cleanly apart. None of those statements removes the predictive power of the standard formalism. The Born rule still works. The Schrödinger equation still describes evolution. What changes is that the mathematics is no longer asked to float free of ontology. There is something the equations are about, and that something is a real physical configuration of a real physical medium.

DR. S: So the gain is not that the probabilities go away.

DR. U: No.

DR. S: Nor that every quantum mystery is finished.

DR. U: No.

DR. S: The gain is that physics is allowed to ask what it is talking about again.

DR. U: Yes. That is the gain. And it is not a small one.

DR. S: And the further gain, if it is real, is that the ontology begins to point toward new control strategies rather than merely new language.

DR. U: Exactly.

DR. S: Which is either a genuine virtue or a new place to fail.

DR. U: As it should be.

That was enough.
Not because the disagreement had been resolved.
Because it had been made exact.

Dr. S did not deny that the standard interpretation had left deep questions hanging for nearly a century.
Dr. U did not deny that replacing interpretive ambiguity with ontology creates new burdens of its own.

Both of those things were true.
Which was why the exchange worked.

DR. S: Let me try the summary test.

DR. U: Go ahead.

DR. S: On your view, the wavefunction is a real ULF field configuration rather than a merely probabilistic device. Particles are stable coherent excitations in that field. Measurement is a physical localization process involving decoherence, irreversible interaction, and mode selection. Entanglement is fundamentally a non-separable ULF structure rather than a pair of independent systems exchanging hidden signals. Configuration space is useful mathematics, not fundamental physical reality. And in the fuller version of the theory, the UCF may serve as a coordination layer during irreversible localization, but not as an energy- or information-carrying superluminal channel.

DR. U: That is fair.

DR. S: And the stronger claim is that this is not only an interpretive cleanup. If quantum states are real resonant field structures, then coherence control should become a more direct engineering problem. Resonant Stabilization would then be one of the first places where the ontology could begin to cash out technologically rather than only philosophically.

DR. U: Exactly.

DR. S: I can see why you think this matters.

DR. U: Yes.

DR. S: Mattering is not the same as being right.

DR. U: Yes.

DR. S: Good. Then we know what has to survive.

INTERLUDE

By the end of the exchange, quantum mechanics no longer looked like a theory that had to remain permanently ontologically evasive.

At least in the UFD picture, the wavefunction had become a real field structure, collapse had become a physical localization event, and entanglement had become the persistence of one extended configuration rather than a mysterious line between independent particles.

That was the first gain.

It did not abolish quantum probability.
It did not eliminate every open question about outcome selection.
And it did not pretend that the more ambitious UCF-related extensions had already been driven to closure.

But it did something important. It reduced the amount of quantum mechanics that had to remain conceptually suspended in midair. The theory was no longer being asked to succeed mathematically while refusing to say what sort of world it described.

The second gain was equally important. If the ontology is real, then it should not end with interpretation. It should open new paths of control. Resonant Stabilization mattered for that reason. It made the quantum discussion more than a philosophical repair. It turned it into a possible engineering frontier — and therefore into a more exposed scientific claim. If quantum coherence is genuinely a resonant field phenomenon, then active resonance-guided stabilization should become a place where the ontology can either gain practical support or lose credibility.

That was a real explanatory gain.

It also came with a real cost.

A field ontology earns nothing merely by being easier to picture. It earns its place only if it preserves the unmatched formal success of quantum mechanics while remaining honest about what has truly been clarified, what has only been narrowed, and what still remains open.

Temporary Verdict

Dr. U succeeded in making the quantum world more physically legible. The wavefunction became a real field structure, collapse became a physical event rather than a primitive axiom, entanglement became one extended configuration rather than a spooky link between separate things, and the ontology opened a plausible technological frontier in resonant stabilization.

Dr. S succeeded in holding the decisive line of pressure. A more intelligible ontology is not yet a better quantum theory. It earns its place only if it remains faithful to the formal success of quantum mechanics, distinguishes clearly between core claims and stronger extensions, and survives the practical and experimental burdens created by its own technological implications.

The next question now pressed in from beneath all the others.

IF THE QUANTUM WORLD CAN BE DESCRIBED IN TERMS OF REAL FIELD STRUCTURE RATHER THAN ABSTRACT PROBABILITY ALONE, THEN THE NEXT BURDEN IS THE DENSEST STRUCTURE IN PHYSICS: WHAT ACTUALLY BINDS THE NUCLEUS?

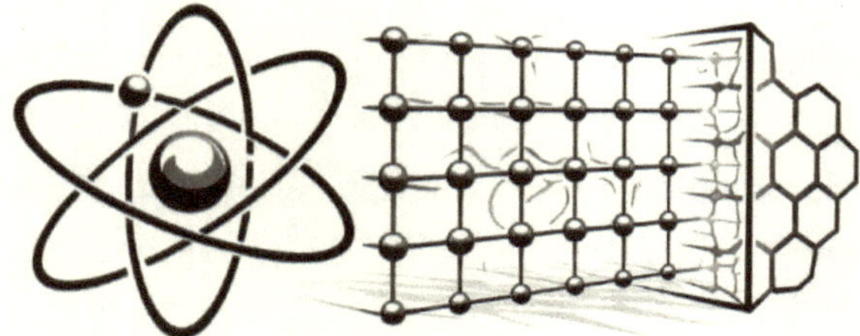

CHAPTER 13

The Geometric Nucleus

The strong force, Helium-3, and coherence under pressure

The nucleus is one of the strangest places in physics. It is so small that for most of human history nobody knew it existed, and so dense that almost the entire mass of every atom — and therefore almost the entire mass of every ordinary object — is concentrated inside it. If an atom were the size of a football stadium, the nucleus would be a marble at the center. And yet that marble carries almost all the weight. Everything else is mostly empty space, lightly occupied by electrons.

The discovery of the nucleus was one of the great shocks of twentieth-century science. Rutherford's scattering experiments showed that the atom was not a diffuse pudding of positive charge, but a structure whose mass was concentrated in a tiny central core. Over the following decades, physicists identified the proton and the neutron and gradually realized that nuclei were tightly bound collections of both. That created a new problem immediately. Positively charged protons should repel one another strongly. Why, then, do nuclei hold together at all? The answer was the strong interaction: a short-range force powerful enough to overcome electrostatic repulsion and bind nuclear matter together.

The standard story today traces that interaction, at its deepest level, to quantum chromodynamics. Protons and neutrons are treated as quark systems bound by gluons, and nuclear binding is then described as a residual effect of that deeper quark-gluon dynamics. The framework is powerful and successful. But it is also notoriously difficult to use in any exact way for real nuclei. In practice, nuclear physics relies on a mixture of deep theory, effective models, and heavy computation. The result is empirically formidable, but the underlying physical picture often remains abstract.

That is why nuclear physics is one of the least forgiving places in the whole book. A theory can sound elegant for quite a while in cosmology, where the data is sparse and the timescales are vast. It can sound elegant for quite a while in foundational interpretation, where the questions are often philosophical. In nuclear physics, elegance has to harden into structure very quickly. The numbers are precise. The measurements are unforgiving. Any alternative picture of the nucleus is exposed almost at once to quantitative pressure.

UFD takes exactly that risk. It offers a much more intuitive picture and therefore places itself under a much sharper burden. On this view, the nucleus is not fundamentally a battleground of irreducible force exchange. It is a geometric and fluid-dynamic system of structured vortices in the dense field. Each proton and neutron is a stable circulation pattern in that field, with its own characteristic shape. Binding happens when those circulation patterns interlock in ways that reduce redundant boundary strain. When two structures that would otherwise each carry their own disturbed field can share part of that burden, the combined system becomes more efficient than the separated parts. The released energy is what UFD calls the Coherence Dividend.

On that picture, the strong force is not primitive. It is the effective name given to what happens when geometric structures in the field settle into lower-strain, more coherent arrangements. UFD calls the underlying tendency the Geometric Coherence Force. The name is meant to emphasize that what standard physics treats as a powerful short-range attraction is, in this view, the field's drive toward more efficient organization of its own circulation patterns.

This is a much more intuitive picture than quantum chromodynamics, and that is precisely why it has to be tested carefully. Helium-3 is one of the first narrow places where the picture must do more than sound appealing. A geometric mechanism for nuclear binding ought to show itself there in a way that is not merely poetic. If it does, the framework has begun to earn credibility in one of the hardest sectors of physics. If it does not, the intuitive appeal counts for nothing.

TECHNICAL NOTE — HELIUM-3 BINDING ENERGY

The intuitive claim is this: nuclear binding energy can be understood as a geometric coherence gain rather than as the residual of a primitive strong force.

More precisely, UFD models the He-3 nucleus as a three-nucleon system whose binding energy arises from the reduction of redundant boundary strain when the nucleon vortices interlock coherently. The calculation uses the geometric docking configuration, the boundary-

strain reduction, and the Coulomb correction for the two like-charged
protons.

The result is a predicted total binding energy of approximately 7.718
MeV, in extremely close agreement with experiment. At the quoted
precision the values nearly coincide, with a residual on the order of
0.002 MeV depending on the exact derived quantity and rounding
convention. This does not close nuclear physics. But it does give the
geometric picture a genuine light-nucleus benchmark in a place where
the structure is still simple enough to count.

Dr. S began with the kind of directness that always meant he was prepared
to be unimpressed.

DR. S: All right. Enough atmosphere. What binds the nucleus?

This was the right opening pressure point.
If UFD could not say clearly what binding is, then everything else in the
chapter would collapse into attractive language.

DR. U: Geometric coherence.

DR. S: Not the strong force.

DR. U: Not as a primitive interaction, no.

DR. S: Then let me say immediately what bothers me. Nuclear physics is
exactly the place where nice physical pictures go to die. Everybody wants a
tangible story about the nucleus. The reason standard physicists become
suspicious is that the nucleus punishes hand-waving very quickly.

DR. U: That is fair.

DR. S: Good.

DR. U: The question is not whether standard nuclear theory has powerful
tools. It plainly does. The question is whether those tools are describing the
deepest cause of binding, or the effective behavior of a more primitive
geometric process.

DR. S: And naturally you think it is the second.

DR. U: I do.

That was the first real edge of the conversation.
Dr. S trusted the mature formal culture around nuclear physics.

Dr. U distrusted any framework that made binding primitive before asking whether binding might instead be the energetic consequence of a deeper structure.

So Dr. S pushed immediately to the physical picture.

DR. S: Fine. Then give me the strong version.

DR. U: A nucleon is a structured vortex in the dense field. Maintaining its boundary costs energy. When nucleons dock coherently, some of that boundary strain becomes redundant. The system no longer has to pay twice for the same exposed interface. The released energy is the binding energy.

DR. S: So binding is not a separate force pulling objects together. It is an energetic gain from more efficient geometry.

DR. U: Exactly.

DR. S: Surface-tension logic.

DR. U: In the broad sense, yes. But not just as metaphor. The point is that the nucleus seeks lower boundary strain.

DR. S: And that tendency is what you call the Geometric Coherence Force.

DR. U: Yes. The force is the drive toward lower-shear coherence. The Coherence Dividend is the energy released when that drive succeeds.

DR. S: So the force and the dividend are not the same thing.

DR. U: Correct. One is the tendency. The other is the payoff.

Dr. S did not interrupt. That usually meant the answer had at least crossed the threshold from vague to attackable.

He pressed at the obvious point.

DR. S: Let me tell you what is attractive and what is dangerous. The attractive part is obvious: this is physically imaginable. The dangerous part is equally obvious: physically imaginable nuclear stories are very easy to tell and notoriously hard to make exact.

DR. U: True.

DR. S: So where does this stop being philosophy and start becoming pressure?

DR. U: Helium-3.

DR. S: Good answer.

DR. U: It has to be. If the geometry means anything, it should show itself in the light-nucleus regime where interface structure is still visible.

That was exactly the right place to go.
The nucleus becomes harder to read geometrically as it grows.
So a geometric theory ought to live or die first where the structures are still relatively simple.

Dr. U did not hesitate.
That was wise.

DR. U: Start with the deuteron as the calibration of one coherent docking interface. Then move to Helium-3. In UFD, Helium-3 forms a triangular docking geometry. Once the coherent interfaces are counted and the Coulomb penalty is subtracted, the proton separation energy comes out essentially right.

DR. S: You are being careful.

DR. U: I should be. Helium-3 is not the final proof of the whole framework. It is a narrow place where the geometry has to do real work.

DR. S: And why Helium-3 rather than, say, uranium?

DR. U: Because by the time nuclei get large, many-body effects and effective descriptions dominate. If the deepest mechanism is geometric, it should be clearest where the number of coherent interfaces is still small enough to count.

DR. S: So the strategy is to start where the shape still shows.

DR. U: Exactly.

That was one of the stronger moments in the exchange.
Not because He-3 solved nuclear physics.
Because it gave the theory the right kind of vulnerability.
A weaker theory would have rushed toward the biggest phenomena first.
A stronger one began where its core mechanism could still be seen.

DR. S: Fine. But if Helium-3 is the test case, I want the real claim in plain language.

DR. U: The real claim is simple. If nucleons are vortices, and if nuclear binding is the energy gained when they fit together with less exposed boundary, then a small nucleus like Helium-3 should show that gain in a way we can estimate geometrically. UFD says it does.

DR. S: And the Coulomb repulsion?

DR. U: Real, but not the main actor. Coulomb repulsion opposes coherence gain. It does not create the binding in the first place. Binding comes from reduced boundary strain. Coulomb cost is then subtracted from that gain.

DR. S: So electrostatic repulsion is the tax, not the source of the income.

DR. U: That is a fair way to put it.

That helped, because it made the logic clearer without making it thinner. Coulomb repulsion explained part of what resists binding.
It did not, on the UFD picture, explain what makes binding possible in the first place.

Dr. S then pressed on the standard objection that had to come next.

Precision.

DR. S: All right. But here is the mainstream response. Light-nucleus fits are interesting. They are not enough. Standard nuclear physics does not live or die by whether one can tell a good story about Helium-3. It lives by an enormous culture of quantitative success.

DR. U: Of course.

DR. S: So what exactly are you claiming to replace?

DR. U: The deepest ontology of binding. Not every effective technique used in nuclear modeling. The core replacement is this: the strong interaction is not primitive. Nuclear binding is a coherence gain from geometric docking. The Geometric Coherence Force is the system's drive toward that lower-shear state, and the Coherence Dividend is the released energy when it gets there.

DR. S: And what remains open?

DR. U: The full many-body closure at larger nuclei, the exact detailed mapping of shell effects from first principles, and the full transport from the geometric picture into every practical nuclear calculation. Those are stronger burdens.

DR. S: Better.

DR. U: It has to be said that way. Otherwise, the picture would overstate itself.

That changed the temperature.

A shallower version would have pretended that a clear mechanism in light nuclei automatically meant total closure everywhere else.

A stronger one knew the difference between a foundational replacement and a finished computational empire.

Dr. S respected that enough to keep going seriously.

DR. S: Let us talk about Helium-4 for a moment, then. Because if your story is really one of geometric stability, there ought to be especially stable shapes.

Helium-4 is one of the most striking examples of unusual stability in nuclear physics. It is the nucleus of ordinary helium, made of two protons and two neutrons, and it is held together more tightly per nucleon than almost any other light nucleus. It appears everywhere: in stellar fusion, in alpha decay, and in the basic architecture of light-element nucleosynthesis. Something about the four-nucleon arrangement is unusually efficient.

DR. U: Yes. In UFD, that is exactly the point. Helium-4 is treated as a coherence maximum — a particularly efficient closed geometry that the field naturally wants to settle into.

DR. S: Tetrahedral.

DR. U: In the strongest version, yes.

Dr. U paused for a moment, because the word he had just confirmed needed to be unpacked for it to mean anything.

DR. U: A tetrahedron is the simplest fully closed three-dimensional shape: four vertices, four faces, every vertex connected to every other by equal distance. In UFD's reading of Helium-4, the four nucleons occupy the vertices of that closure, and their circulation patterns interlock in the tightest geometric arrangement four objects can achieve in three dimensions. There is no leftover asymmetry and very little exposed boundary. The configuration is closed in a way that almost no other small nucleus can match.

DR. S: So the more geometrically complete the docking, the greater the energetic payoff.

DR. U: Exactly. Stability and symmetry are not separate facts in this picture. They are different descriptions of the same coherence gain. Helium-4 is special because it is the smallest nucleus that achieves a fully closed three-dimensional arrangement.

For a moment the conversation quieted.
The nucleus was no longer only a prison of protons and neutrons held together by an abstract force.
It had become a search for the most efficient way to close boundary strain — for the geometric configurations that let circulation patterns in the field share their burdens most completely.

But Dr. S was not about to let the gain stand alone.
He pushed toward implication.

DR. S: Fine. Suppose I grant that this is a coherent re-reading of nuclear binding. What does it buy you besides a cleaner story?

DR. U: At minimum, a different research program.

DR. S: More specifically.

DR. U: If nuclear binding is partly geometric — if the barriers between configurations depend on how well two structures can dock in the field — then some transformations may be guided not just by hitting nuclei harder, but by shaping the field environment around them so that the docking becomes easier.

DR. S: Fusion.

DR. U: Potentially. And perhaps transmutation as well. But the near-term claim has to be smaller and cleaner. I am not promising cheap fusion next year. I am saying that if nuclear barriers are partly geometric, then there should be measurable, reproducible effects under deliberately shaped field conditions: modest barrier lowering, pathway redirection, or small but consistent shifts in reaction rates that the standard energetic picture would not naturally predict.

DR. S: So again the right posture is staged.

DR. U: Exactly. If the ontology is right, Geometric Catalysis should first show up as small but reproducible laboratory effects, not as science fiction on demand. The first generation of experiments would not produce abundant energy. They would produce statistical signatures — slightly higher reaction rates than expected under conditions where the field geometry has been shaped in specific ways. If those signatures appear and survive scrutiny, the framework has begun to earn the right to be taken seriously as a guide to nuclear engineering. If they do not appear, the geometric reading remains a philosophical preference with no practical traction.

That improved the conversation.
It showed that the nuclear picture was not only explanatory.
It was also, at least in principle, technologically generative.
And because Dr. U kept the claim modest, it strengthened rather than weakened the exchange.

Dr. S then widened the frame one final time.

DR. S: Let me tell you what I think is strongest here. You are replacing the strong force with one geometric postulate rather than a patchwork of separate ingredients. Binding becomes lower boundary strain. The force becomes the drive toward coherence. The energy becomes the dividend released when the geometry succeeds. Helium-3 becomes a narrow pressure point, and even the first technological horizon follows from the same logic.

DR. U: Yes.

DR. S: That is a real unification if it works.

DR. U: Exactly.

DR. S: The danger is obvious too. Nuclear physics is not forgiving. A vivid picture earns nothing unless it tightens into numbers, structure, and eventually control.

DR. U: Agreed.

DR. S: In other words: surface-tension language is not enough.

DR. U: Never enough.

That was the center of gravity.
Dr. U was not merely offering a more intuitive picture of the nucleus.
He was claiming that one geometric mechanism could replace the deepest causal story of binding.
Dr. S was not merely defending orthodoxy.
He was insisting that replacement-level ambition must survive the exact terrain where nuclear physics is least charitable.

Nothing about that was settled.
But the burden had become exact.

DR. S: Let me try the summary test.

DR. U: Go ahead.

DR. S: On your view, nuclear binding is not fundamentally caused by a primitive strong force. It is the energy gained when structured nucleon vortices dock in ways that reduce redundant boundary strain in the dense field. The Geometric Coherence Force is the drive toward that lower-shear state, and the Coherence Dividend is the released binding energy when it is achieved. Helium-3 matters because it is a light-nucleus case where the coherent interfaces are still simple enough to count geometrically, and Helium-4 matters because greater symmetry yields greater coherence gain. Coulomb repulsion remains real but secondary: it opposes the gain rather than creating the binding. And if this picture is right, then even the first technological implication follows naturally — not miracle engineering, but the possibility that barrier crossing may be guided by field-shaped geometric conditions rather than brute forcing alone.

DR. U: That is fair.

DR. S: The attraction is obvious.

DR. U: Yes.

DR. S: The danger is equally obvious.

DR. U: Yes.

DR. S: Good. Then the burden is finally clear.

INTERLUDE

By the end of the exchange, the nucleus no longer looked like a place where one simply stopped asking for a physical picture.

At least in the UFD view, it had become the clearest place where geometry, pressure, and coherence were supposed to cash out as binding.

That was the gain.

The nucleus was no longer a battleground of irreducible exchange first and an intuitive story second. It had become a system seeking lower exposed boundary, with the strong interaction re-read as the effective name for that tendency. The light nuclei mattered because they made that claim vulnerable in exactly the right way. Helium-3 was narrow enough to test the counting logic. Helium-4 was symmetric enough to show why closure matters. And even the first technological implication — Geometric Catalysis — followed naturally once barriers were treated as partly geometric rather than merely energetic.

That was a real explanatory gain.

It also came with a real cost.

A vivid picture of the nucleus is worth nothing unless it keeps tightening under quantitative pressure. If the geometry cannot continue to organize the light-nucleus regime, then the coherence story remains only a story. If it can, then one of the oldest mysteries in physics may turn out to have been less about hidden particles than about hidden shape.

Temporary Verdict

Dr. U succeeded in making the nucleus physically legible. Binding became geometric coherence gain, the strong force became an effective name for the drive toward lower-strain form, Helium-3 became a real narrow test case, and Geometric Catalysis appeared as a plausible downstream consequence rather than a detached technological fantasy.

Dr. S succeeded in holding the decisive line of pressure. Nuclear physics is too exacting to be won by vividness alone. The geometric picture earns its place only if it keeps converting intuitive coherence into disciplined structural and quantitative constraint.

The next question now followed naturally.

If geometry really governs the nucleus, then the next step is to ask whether that geometry reaches outward into the atom and the structure of chemistry itself.

Chapter 14

The Resonant Atom

The Platonic nucleus, orbitals, and the coherence landscape

Chemistry is one of the most successful sciences ever developed. It explains why salt dissolves in water, why iron rusts, why oxygen combines with hydrogen to form water but helium does not easily combine with anything at all, and why the enormous variety of substances around us behave the way they do. It does this through a framework that took centuries to assemble and that, by the early twentieth century, had become remarkably powerful.

The structure of the atom lies at the center of that success. In the nineteenth century, Dalton revived the atomic theory of matter and gave it quantitative form. Mendeleev's periodic table then revealed that the chemical elements were not a random list but an ordered system with recurring patterns. In the early twentieth century, Thomson discovered the electron, Rutherford revealed the nucleus, and Bohr connected atomic structure to discrete energy levels. By the 1920s, quantum mechanics had replaced Bohr's little planetary orbits with orbitals — structured regions associated with allowed electron states. From that framework came the modern explanation of the periodic table, chemical bonding, and the broad architecture of chemistry itself.

That framework is one of the great triumphs of modern science. Chemical behavior is explained through electron configuration, shell structure, symmetry, and energy minimization. Why sodium reacts so easily, why neon is inert, why carbon forms four bonds, why the periodic table repeats in regular ways — all of that follows, at least in principle, from the rules governing how electrons occupy orbitals around a nucleus. Any alternative framework that touches chemistry has to preserve what this picture gets right.

In the standard view, however, the nucleus is mostly chemically passive. What matters first is its total charge, because that sets the number of electrons and therefore the orbital structure. Differences in nuclear mass and spin can matter in subtler ways, but the internal geometric arrangement of the nucleus is usually treated as chemically secondary. For most purposes, the deep architecture of the atom is electron architecture.

UFD does not propose to overturn that framework. It proposes to thicken the ontology underneath it.

In the UFD picture, the atom is not a featureless point nucleus surrounded by a mathematically useful probability cloud. It is a coupled resonant system: a structured nucleus surrounded by real standing-wave patterns in the field that fills space. The orbitals described by quantum chemistry are, on this view, not merely probability distributions. They are real three-dimensional resonant patterns in the medium. The electron is not a little point smeared into uncertainty. It is a stable field configuration, and the orbital is the shape of that configuration around the nucleus. The standard orbital mathematics is retained. What changes is what those mathematical structures are taken to describe.

The larger departure concerns the nucleus itself. UFD proposes that nuclei tend toward highly symmetric, low-strain closure patterns — geometric configurations that minimize the strain they impose on the surrounding field. On this view, some nuclei are more stable than others not only because of shell effects or nucleon counts, but because some geometric arrangements are more efficient than others. UFD tracks one aspect of this through the Atomic Stability Index (ASI), which is meant to measure how closely a nucleus approaches an ideal low-strain closure geometry. The ASI is not a complete account of nuclear stability, but it is one way of capturing something the standard picture usually leaves in the background: nuclei have shapes, some shapes are more coherent than others, and those differences may matter outwardly.

The result is a picture in which the atom becomes physically legible from the inside out. The nucleus is no longer just a charge source. It is a structured object with its own geometry. The electrons around it are no longer merely probabilistic bookkeeping devices. They are real standing-wave patterns coupled to that geometry through the same medium. Orbitals become physical resonances of the combined system. Chemistry then becomes, on the UFD reading, the behavior of a real coupled structure rather than a calculation about an abstract one.

Maturity note. The resonant-atom picture is a physically thickened ontology of the atom rather than a quantitative replacement for standard quantum chemistry. Standard orbital calculations and shell-filling rules are preserved. What changes is the physical interpretation of what orbitals are and why certain nuclear geometries may favor particular chemical behaviors. The quantitative claims in this chapter are constrained by the framework but are not yet fully first-principles closed.

Dr. S opened with the kind of skepticism that only comes from a subject already crowned with success.

DR. S: All right. We have reached chemistry, which means we have reached a place where standard theory already works beautifully. So tell me why I should want a new ontology for the atom at all.

DR. U: Because the current one still leaves the atom strangely ghostlike.

DR. S: Ghostlike.

DR. U: Yes. Standard chemistry predicts astonishingly well. But when you ask what an orbital physically is, the answer becomes evasive: a probability distribution, a solution to an equation, a mathematical object in an abstract formalism. Useful, yes. Tangible, not really.

DR. S: That is the old complaint again. "You calculate well but do not picture enough."

DR. U: Not quite. The complaint is not that the picture is unfriendly. It is that the ontology is thinner than the physics seems to invite.

DR. S: And your alternative?

DR. U: The atom is a resonant system. The nucleus is structured. The orbitals are real standing waves. The rules of chemistry follow from the way those standing waves can and cannot organize around that structure.

That was the first hinge of the exchange.
Standard chemistry did not need a richer ontology in order to calculate.
UFD did not deny that.
It claimed instead that the success of the calculations had hidden a physically thicker system than the standard account was willing to describe.

Dr. S pushed directly into the atom itself.

DR. S: Fine. Then tell me the strong version. What is an atom?

DR. U: A coupled two-field structure. At the center is a structured nucleus belonging to the dense field layer that gives nuclei their internal geometry. Around it form stable standing waves in the lighter field layer that carries light and electromagnetic phenomena. Those standing waves are what we call orbitals.

DR. S: So the nucleus is not just a positive point charge generating a potential well.

DR. U: Not on this view. The nucleus is a structured resonant core, and that structure matters.

DR. S: Matters how?

DR. U: It helps determine the stable harmonics available around it. The easiest analogy is a bell. The shape of the bell constrains the tones it can sustain. A round bell rings differently from an oblong one, and a cracked bell rings differently from an intact one, because the geometry of the object determines which standing-wave patterns can stably exist within it. The same principle applies to atoms. The geometry of the nucleus constrains the standing-wave patterns that the surrounding field can form around it.

DR. S: So the orbital is not just a probability cloud around a featureless center.

DR. U: Exactly. It is a real standing-wave mode around a structured core.

Dr. S let that stand for a moment. Then he raised the obvious objection. Not the standard one. The internal one.

DR. S: All right. But before we go further, I want a distinction on the table. There have been other attempts to make the atom more visual or more geometric. Some models emphasize alpha clustering in the nucleus. Others try to make atomic architecture more tangible by giving the atom a more explicit internal layout. Why is this not just another one of those?

DR. U: Because UFD is not merely making the nucleus prettier.

DR. S: More specifically.

DR. U: The alpha-cluster picture, at its best, is mainly a nuclear picture. It says something about how certain nuclei may organize internally. UFD agrees that alpha-like coherence matters deeply in many nuclei. But it goes further. It says that nuclear geometry is not only an internal nuclear fact. It shapes the surrounding orbital landscape.

DR. S: And those more explicitly structured atom models?

DR. U: Same answer in another form. UFD is not proposing a visually appealing rearrangement of atomic parts. It is proposing a coupled-field atom in which the nucleus and the orbitals belong to one resonant system. The geometry of the dense-field core and the harmonics of the lighter-field shell are not separate stories. They are one architecture.

DR. S: So the gain is not merely that the atom becomes more imaginable.

DR. U: No. The gain is that the nucleus stops being chemically mute and the orbitals stop being ontologically thin.

That mattered. Because Dr. S had identified the immediate danger correctly. A theory like this could easily sound like one more attempt to make the atom feel less abstract. Its burden was to show that it was doing more than that. So he pushed where the real pressure belonged.

The nucleus.

DR. S: Fine. Then let us begin at the center. What do you mean by a Platonic nucleus?

DR. U: I mean that nuclei tend toward highly symmetric, low-strain closure patterns. Not every nucleus achieves them equally well, but the most stable ones are not arbitrary piles. They approach clean geometric arrangements — configurations in which the internal structure achieves the most efficient possible closure of the field's circulation patterns.

DR. S: And the Atomic Stability Index, this number you keep mentioning, is supposed to measure that?

DR. U: Carefully. The Atomic Stability Index, or ASI, tracks one important dimension of it. It is not a full law of stability by itself.

DR. S: Good. Say that more loudly.

DR. U: Happily. ASI is a diagnostic, not a complete theory. It tells you how well a nucleus approaches closed, alpha-like geometric completion. But the broader question of nuclear and atomic stability belongs to a larger picture.

DR. S: Meaning?

DR. U: Meaning that symmetry matters, shell closure matters, geometry matters, coupling matters, and in the atom the orbital side matters too. You should not confuse one useful stability index with the full physical reality it helps describe.

That was one of the most important clarifications in the conversation. A weaker version of the theory would have tried to make the ASI do everything.

A stronger version knew its proper role.

Dr. S respected that immediately.

Which was why he pressed forward rather than dismissing the point.

DR. S: Good. Then what exactly is the larger picture?

DR. U: The full space of stability conditions. Think of it as a kind of landscape — a topography in which elevation represents how stable a configuration is. ASI tells you something important about one feature of that landscape: how well the nucleus closes geometrically. But a real atom is not just a nucleus. It is a nucleus plus a field of standing-wave possibilities around it. So the actual stability of the atom depends on how well those two sides fit together.

DR. S: So a geometrically elegant nucleus is not enough by itself.

DR. U: Correct. The atom is stable when the structured core and the standing-wave shell reinforce one another. A nucleus may be highly coherent in itself and still produce a different atomic landscape than another equally closed nucleus, depending on how the surrounding field can organize around it.

DR. S: Which means the chemistry cannot be read off from one number alone.

DR. U: Exactly. The full coherence landscape is richer than ASI. ASI is one sign inside that landscape, not the landscape itself.

The conversation had now shifted from nuclear stability in itself to atomic stability as a coupled-field problem. That was the right shift. Because the real burden of the resonant atom was not merely that nuclei had internal geometry, but that this geometry had to matter outwardly — to the surrounding standing-wave system that made the atom chemically active.

Dr. S pressed directly on that point.

DR. S: Fine. Then tell me the strongest version of the claim. Why should the nucleus matter chemically at all?

DR. U: Because it sets the harmonic boundary conditions.

DR. S: Meaning?

DR. U: The orbitals are not free-floating mathematical possibilities. They are stable standing-wave modes in a real medium, and standing-wave modes depend on the structure of what they form around. Change the shape, and the allowed tones change with it. If the nucleus is a structured core rather than a featureless point, then its geometry helps determine which orbital harmonics are stable, how they orient in space, and how they are spaced in energy.

DR. S: So the nucleus is shaping the atomic landscape directly.

DR. U: Yes.

DR. S: That is a very strong claim.

DR. U: It has to be. Otherwise, the structured nucleus is chemically irrelevant, and the whole inner half of the ontology goes silent at the very place where the outer half begins.

That was the center of the model's burden.
If the nucleus did not shape the surrounding standing-wave system in a physically meaningful way, then the resonant atom would collapse back into a more decorative version of standard chemistry.

Dr. S heard that clearly and pressed where the claim had to cash out.

DR. S: All right. Then what becomes of the standard orbitals?

DR. U: They remain, but their status changes.

DR. S: Go on.

DR. U: In UFD, the familiar orbital shapes — s, p, d, and f — are not abstract solutions first and physical objects second. They are the stable overtone patterns of the field around the nucleus, the way the resonant modes of a vibrating drumhead are real physical patterns rather than mathematical abstractions.

DR. S: So the Schrödinger solutions survive.

DR. U: Yes. But they are being read as the harmonic mode structure of a real field. The mathematics stays. What changes is what the mathematics is mathematics of.

DR. S: And the orbital energies?

DR. U: The energies of those stable modes, shaped by the field environment and by the geometry of the core they surround.

Dr. S did not object immediately.
That was one of the clearest moves available to UFD in this domain: do not

discard the orbital mathematics; say what the mathematics is mathematics of.

So he pushed next into the rules that chemistry students are taught to memorize.

DR. S: Fine. Then what becomes of the rules of chemistry? Shell filling. Exclusion. Periodicity. Are these still axioms, or do you think you have physical reasons for them?

DR. U: Physical reasons.

DR. S: Start with exclusion.

DR. U: In this picture, the Pauli exclusion principle becomes a stability law of resonant occupancy. Two identical standing-wave excitations cannot occupy the same stable state without destabilizing the field organization that defines that state in the first place.

DR. S: So Pauli becomes a consequence of what stable resonance requires rather than a brute rule.

DR. U: Exactly.

DR. S: And shell filling?

DR. U: The atom fills outward through the stable harmonic hierarchy of the field around the nucleus. Lower modes stabilize first. Higher modes appear only when the lower ones are occupied and the field geometry permits them. The order of filling is not arbitrary. It reflects which standing-wave patterns can stably coexist around a given nuclear core.

DR. S: So again the rules become consequences of what the medium can stably hold.

DR. U: Yes. That is the recurring logic here.

That clarified the ambition of the theory.

The goal was not to rename chemistry.

It was to replace axioms with mechanics wherever possible.

That is always a dangerous move.

Which is why Dr. S now pushed on the broadest chemical structure of all.

The Periodic Table.

DR. S: All right. Then what is the Periodic Table in your view?

DR. U: A map of stable harmonic architectures.

DR. S: Explain that without poetry.

DR. U: The periodic recurrence of chemical behavior reflects recurring patterns in the outermost electron configurations — what chemists call valence patterns. Standard chemistry already knows this and explains it through orbital filling. UFD keeps that explanation. But UFD adds that those valence patterns are not floating above the nucleus. They are the outer resonant expression of a coupled inner geometry. So the Periodic Table is not just a record of electron bookkeeping. It is the visible ordering of stable atom-wide resonances.

DR. S: So elements differ not only because they have different nuclear charges, but because they have different coupled nucleus-and-wave architectures.

DR. U: Exactly.

DR. S: Which means periodicity is deeper than shell counting.

DR. U: It is shell counting physically understood.

That was one of the better turns in the conversation.
It made the point without exaggerating it.
UFD was not denying periodicity as standard chemistry understood it.
It was claiming that periodicity could be given a more physically grounded basis.

Dr. S then turned, as he should have, to evidence.

DR. S: Fine. Then where should this picture show up in ways standard chemistry does not already expect?

DR. U: In subtle effects that depend on nuclear geometry.

DR. S: Such as?

DR. U: Three main places. First, orbital asymmetries around non-spherical nuclei. Second, geometry-sensitive responses to external electric fields — especially in the fine structure of Stark splitting. Third, once we move to bonding, small but measurable bond-energy anomalies that track nuclear coherence rather than electron shell structure alone.

DR. S: So we are already leaning toward the next conversation.

DR. U: Inevitably. Once the atom is resonant, coupling between atoms becomes the next burden. But even before that, the atom itself should reveal traces that a purely featureless nucleus picture would tend to smooth away.

That mattered.

Because it kept the conversation from floating off into ontology alone.

A resonant atom had to leave traces.

Otherwise it was only a more satisfying picture layered on top of standard results.

Dr. S then brought the pressure back to the question he had been carrying from the start. Why believe this at all?

DR. S: Let me tell you what I think is strongest here. You are not merely saying orbitals are real. You are tying them to nuclear geometry. That makes the atom thicker, more physical, and more unified. The Platonic nucleus matters. The ASI matters. But neither is being allowed to pretend to be the whole story, because the full stability question belongs to a wider coherence landscape. That is all to your credit.

DR. U: Thank you.

DR. S: Do not thank me yet. The danger is obvious too. Chemistry is one of the most successful sciences in history. The moment you tell me the nucleus matters more than standard chemistry thinks, you inherit a very large burden.

DR. U: I know.

DR. S: Good.

DR. U: And the burden is not vague. If the resonant atom is right, then the structured nucleus should leave detectable traces in orbital shape, level ordering, symmetry response, and later in bonding. If those traces never appear, then the extra ontology becomes hard to justify.

DR. S: Better.

DR. U: It has to be said that way.

That was the center of gravity.

Dr. U was not merely offering a prettier picture of the atom.

He was claiming that one unified field architecture reached all the way from nuclear geometry into the visible rules of chemistry.

Dr. S was not merely defending orthodoxy.

He was insisting that the most successful subject in this part of science does not surrender its ontology cheaply.

Nothing about that was settled.
But the burden had become exact.

DR. S: Let me try the summary test.

DR. U: Go ahead.

DR. S: On your view, the atom is a coupled resonant system: a structured UEF nucleus surrounded by real ULF standing waves. The nucleus is not just a point source of charge, but a geometric core whose closure patterns matter. ASI helps diagnose one part of that coherence, but broader stability belongs to a larger coherence landscape rather than to one number alone. Orbitals are real harmonic modes, not merely probability clouds. The standard rules of chemistry — exclusion, shell filling, and periodic recurrence — are therefore not brute axioms, but consequences of what stable coupled field structures can and cannot support. And the Periodic Table becomes a map of recurring atom-wide resonant architectures rather than only a bookkeeping scheme for electron occupancy.

DR. U: That is fair.

DR. S: That is a cleaner picture.

DR. U: Yes.

DR. S: A cleaner picture must still survive a dirtier world.

DR. U: Yes.

DR. S: Good. Then the pressure is where it should be.

INTERLUDE

By the end of the exchange, the atom no longer looked like a featureless charge center surrounded by an ontologically ambiguous cloud.

At least in the UFD picture, it had become a coupled resonant system whose inner and outer structures belonged to one field architecture. The nucleus was no longer chemically mute. Orbitals were no longer merely mathematical envelopes. Shell rules were no longer just accepted constraints. And the Periodic Table was no longer only a remarkably effective ordering chart. All of them had been drawn into one physical picture.

That was the gain.

It was also the limit of what had been established here.

Because a resonant atom is not yet chemistry in full. It is the precondition for chemistry. The deeper burden begins when two such systems meet: when their standing waves overlap, when their nuclear coherence either aligns or frustrates, and when bond strength starts to depend not only on electron bookkeeping, but on congruence across the full resonant architecture.

That is where the next conversation begins.

TEMPORARY VERDICT

Dr. U succeeded in making the atom physically legible. The structured nucleus, the real standing-wave orbitals, ASI as a diagnostic, and the broader coherence landscape all came together as one coupled picture rather than a collection of disconnected ideas.

Dr. S succeeded in forcing the distinctions that matter most. A diagnostic is not a full law, a more visual atom is not yet a better atom, and a richer ontology earns its place only if the structured core leaves observable traces in the orbital behavior it claims to shape.

The next burden now follows naturally.
If the atom is truly resonant, then chemistry begins when resonant atoms try to fit together.

That is where geometric congruence, RDI, and the bond itself come into view.

CHAPTER 15

The Resonant Bond

Geometric congruence, RDI, and the physical basis of chemistry

Chemical bonds are what hold the world together. Every solid object you have ever touched, every drop of water, every breath of air, every cell in your body, exists because atoms can join with other atoms to form stable structures. Without chemical bonds, there would be no molecules, no materials, no chemistry, and no life. The question of what a bond actually is — what physically holds two atoms together once they remain together — is therefore one of the most important questions in all of science.

Modern chemistry answers that question with extraordinary success. Covalent bonds are described through shared electron density, ionic bonds through charge transfer and electrostatic attraction, and metallic bonds through delocalized electron states spread across a lattice. Those frameworks work remarkably well. They predict molecular shapes, bond energies, spectroscopic behavior, crystal structure, and a vast range of chemical phenomena with practical accuracy. Computational chemistry is already powerful enough to guide real laboratory work. Any alternative picture of bonding has to preserve that success.

And yet the physical picture can still feel thinner than the mathematics. A covalent bond is said to be a region of shared electron density that lowers the total energy of the system. That is correct, and it predicts what it needs to predict. But it can still leave the bond itself feeling less like a physical event than like a fortunate feature of a successful calculation. Chemistry knows how bonds behave. It is less explicit about what bond stability physically consists in.

That is where UFD enters. It does not propose to abandon the standard machinery of quantum chemistry. It proposes to thicken the ontology underneath it.

In the UFD picture, a chemical bond is not, in the deepest sense, just the sharing or transfer of electrons. It is the formation of a more stable joint resonance between two atoms. This claim depends directly on the previous chapter's view of the atom. If an atom is a coupled resonant system — a structured nucleus surrounded by real standing waves in a physical medium

— then bonding is what happens when two such systems come close enough for those standing-wave structures to merge into one larger, lower-strain resonant pattern.

The stability of that merged pattern depends on two levels at once. One is orbital overlap: how well the outer standing-wave structures of the two atoms can merge into one coherent mode. Standard chemistry already knows that this matters. UFD agrees. The second is nuclear congruence: how well the nuclei serve as a geometric foundation for the combined standing wave. UFD's added claim is that the nuclear contribution is not always negligible. The bond is a property of the whole coupled system, not only of the electron clouds.

The extra stability from a successful merger is what UFD calls the Resonance Dividend. It is the energy released when two resonant systems find that they can sustain themselves more efficiently together than apart. The Resonance Dividend Index, or RDI, is not a complete theory of bonding. It is a diagnostic — a way of tracking whether a bond realizes slightly more coherence or slightly less than standard quantum chemistry alone would predict. In that sense, RDI plays a role analogous to the ASI in the previous chapter: it is one constrained sign inside a broader picture, not the whole picture by itself.

The result is a thicker picture of chemistry. A bond is no longer only a successful calculation about shared electrons. It is a real physical event in which two resonant systems discover that they can persist more efficiently as one combined structure than as two separate ones. The standard mathematics is preserved. The standard energies and structures are preserved. What changes is what the mathematics is taken to describe.

Maturity note. The resonant-bond picture and the Resonance Dividend Index are presented as a physically deeper reading of chemical bonding, not as a completed alternative to computational quantum chemistry. Standard bond energies, molecular geometries, and spectroscopic predictions are preserved. The burden of producing independent quantitative predictions remains open.

The previous chapter argued that an atom is a coupled resonant system rather than a point nucleus surrounded by an abstract cloud. Once that picture is in place, the next question follows naturally: what happens when two such systems come close enough to couple?

Dr. S began exactly where he should have.

Dr. S: Fine. We have spent a whole conversation making the atom thicker. Now the obvious question follows. What is a chemical bond on your view?

Dr. U: A merged resonance.

Dr. S: Meaning?

Dr. U: Two atoms do not merely sit near one another while their electrons happen to arrange themselves favorably. They form a new joint standing-wave system. The bond is the stable merged mode that results.

Dr. S: So not bookkeeping first, physical union second.

Dr. U: Exactly. The bookkeeping remains useful. But underneath it, the bond is a real coherence gain — a real physical event in which two resonant systems discover that they can sustain themselves more efficiently as one larger system than as two separate ones.

Dr. S: And that coherence gain is what you call the Resonance Dividend.

Dr. U: Yes. The system settles into a more stable geometry and releases energy by doing so. The released energy is the dividend the field pays out for finding a more efficient arrangement.

That immediately clarified the ambition of the model.
The goal was not to discard chemistry.
It was to say what chemical stability was stability of.

Dr. S pressed where he had to next.

Dr. S: All right. Then I want the distinction made early. Are you claiming to replace quantum chemistry, or to sit beneath it?

Dr. U: Beneath it.

This mattered because it kept the theory disciplined. UFD was not claiming to discard quantum chemistry's baseline bond calculations. It was claiming that those calculations might leave out a small but real coherence contribution tied to nuclear geometry.

Dr. S: Good.

Dr. U: The baseline bond energies, electronic structure, and chemical specificity remain where standard quantum chemistry already handles them well. UFD is not trying to erase any of that. The stronger claim is narrower. If the broader coherence picture is right, then nuclear geometry contributes a small additional modifier to how fully a bond realizes its baseline stability. That is where the Resonance Dividend Index comes in.

Dr. S: So again, you are not saying the baseline theory is useless.

Dr. U: No. I am saying it may be incomplete in a specific and testable way. That was worth lingering on.
A weaker theory would have tried to overthrow chemistry in one dramatic sentence.
A stronger one knew exactly where its claim began and where it did not.

Dr. S respected that enough to keep pressing seriously.

Dr. S: Fine. Then tell me what the Resonance Dividend Index actually is.

Dr. U: A diagnostic of residual coherence gain or frustration.

Dr. S: That is not yet clear enough.

Dr. U: Let me try it this way. Standard quantum chemistry predicts how stable a bond should be from electron arrangement alone. That gives you a baseline — the bond energy you expect, the vibrational frequency you expect, the geometry you expect. UFD then asks whether the real bond is slightly more stable, slightly less stable, or exactly as stable as that baseline predicts. The Resonance Dividend Index, or RDI, is the comparison between the two.

Dr. S: So positive means better-than-baseline coherence.

Dr. U: Yes. Positive RDI means extra coherence beyond what the electron structure alone would suggest. Negative RDI means geometric frustration — the bond is doing slightly worse than the baseline would imply. Zero means the standard picture is already complete for that case.

Dr. S: And this is a diagnostic, not a law.

Dr. U: Exactly. That word matters here just as much as it did with the Atomic Stability Index. RDI is not, in its present form, a universal derived law. It is a way of asking whether the molecular extension of the coherence picture is showing up at all.

This was the model's most important maturity line.
RDI was not being presented as a universal law of bonding.
It was a diagnostic — a way of asking whether the proposed extension was visible anywhere at all.

So Dr. S pushed on the deeper structure underneath it.

Dr. S: All right. Then what actually determines whether a bond gains this extra coherence?

DR. U: Two things at once. Nuclear congruence and orbital overlap.

The model's core claim was now fully visible.

Molecular stability was being treated as a two-level problem: inner nuclear congruence and outer orbital overlap.

UFD only survives here if those two levels really interact in a disciplined way.

DR. S: Start with the first.

DR. U: Nuclear congruence means the geometric compatibility of the underlying nuclei. If two atoms have nuclei whose structured cores provide a more harmonious foundation for the bond — nuclei whose internal arrangements fit together cleanly rather than clashing — then the merged system can settle more completely into its lowest-strain state.

DR. S: And orbital overlap?

DR. U: The stability of the outer standing waves when they merge. Even if the nuclei fit well, the bond still has to be realized in the electronic field around them. So the actual molecular stability is a two-level problem: inner geometric fit between the nuclei and outer resonant overlap between the standing waves.

DR. S: So the nucleus does not set the whole energy scale.

DR. U: No. That would be too strong. The baseline energy scale remains dominated by the orbital and electronic side, where standard chemistry already does excellent work. Nuclear geometry contributes only as a small modifier, if the extension is right.

The bond had become thicker than standard chemistry usually described it, but it had not become magical.

The claim was still constrained.

Dr. S then turned to the most familiar bond first.

Covalence.

DR. S: Fine. Start with the ordinary case. What is a covalent bond?

DR. U: The shared-resonance case. Two atoms form a joint standing-wave system in which their valence structure — their outermost electron patterns — merges into one stable molecular mode.

DR. S: So shared electrons becomes shared standing-wave structure.

Dr. U: Exactly. The standard language survives because it tracks the same effect. When chemists say two atoms share a pair of electrons, what they are describing is real, and the calculations that follow from that picture are accurate. The deeper picture in UFD is that the bond is a stable merged resonance in the field, anchored at both nuclei.

Dr. S: And what physically makes it stable?

Dr. U: Lower total strain. The merged resonance is more coherent than the two separated atomic resonances would be, so energy is released when the bond forms. That released energy is the Resonance Dividend.

Dr. S: So a covalent bond is the ideal case of shared coherence.

Dr. U: In many ways, yes. It is the cleanest example of two standing-wave systems finding a lower joint mode by merging directly into one another.

That was the obvious place to begin.
But nature does not always choose sharing as its most stable option.

Dr. S knew that and pushed immediately to the next case.
Ionicity.

Dr. S: All right. Then what about the ionic bond? Because there the whole point is that one atom does not share. It gives up.

Dr. U: Exactly. And UFD keeps that distinction.

Dr. S: Then what changes?

Dr. U: The physical reading. In the ionic case, the system lowers its energy by relocating a loosely held resonant structure from one atom into a much more stable unoccupied resonance of another atom. Once that transfer occurs, one side becomes source-heavy and the other sink-heavy. The resulting flow between source and sink in the surrounding medium then holds them together.

Dr. S: So ionic bonding becomes transfer of resonant stability followed by source–sink attraction.

Dr. U: Exactly.

Dr. S: Which means the electrostatic attraction remains real.

Dr. U: Completely real. Sodium chloride still holds together because the positively charged sodium ion is attracted to the negatively charged chloride ion. UFD does not deny that. What it adds is the deeper event underneath.

The transfer happened in the first place because the system found a lower resonant arrangement by relocation rather than by sharing.

This answer mattered because UFD was not flattening all bond types into one vague coherence language. It was preserving the real distinctions among bond forms while re-reading them as different strategies for coherence gain.

Dr. S then turned to the most collective case of all.

The metallic bond.

DR. S: Fine. Then what about metals?

DR. U: The delocalized case.

DR. S: Meaning?

DR. U: Instead of forming one bond here and one bond there, the valence standing waves of many atoms overlap so extensively that they become shared across the entire metallic lattice. Standard theory describes this as a sea of free electrons moving through a lattice of positive ions. UFD describes it as a sea of shared resonance in the field, spread across the whole crystal at once.

DR. S: So the conduction band becomes—

DR. U: A crystal-wide resonant field. The same standing-wave logic that produces a single bond when two atoms come together produces a continuous resonant network when many atoms come together in the right arrangement.

DR. S: Which means metallic bonding is not just many weak local bonds added together.

DR. U: No. It is a collective coherence state.

This was where chemistry began to lean toward condensed matter. Metallic bonding was no longer just another bond type. It was the first large-scale example of coherence spreading beyond local pairwise structure into a collective state. In UFD, covalent, ionic, and metallic bonding had become three strategies by which matter seeks lower-energy coherence: sharing, transfer, and delocalization.

Dr. S then moved to the model's narrowest and most dangerous point. Carbon.

DR. S: All right. Let me guess the real test case. Carbon.

Dr. U: Naturally.

Dr. S: Then make the case.

Carbon was the natural test case for two reasons. First, because carbon is chemically central. Second, because carbon-12 and carbon-13 have been studied with extraordinary precision. Standard chemistry treats them as electronically identical, with only small mass-dependent differences once ordinary isotope effects are accounted for. That makes them an unusually clean place to ask whether nuclear geometry reaches outward into chemistry at all.

Dr. U: Because carbon-12 is the cleanest place where nuclear coherence and chemical importance meet. On the UFD view, its highly symmetric nuclear geometry should not only make the nucleus itself unusually stable. It should also subtly help its bonds achieve a more coherent foundation than carbon-13 bonds can achieve.

Dr. S: And carbon-13 spoils that symmetry.

Dr. U: Exactly. One extra neutron perturbs the geometric closure of the nucleus without changing its electronic identity in any way standard chemistry would notice. From the standpoint of electron configuration, carbon-13 looks just like carbon-12. From the standpoint of nuclear geometry, the two are different.

Dr. S: Which makes it a clean test.

Dr. U: That is the idea. The standard mass effects can be calculated and corrected for with great precision. Whatever is left after that correction is either zero, in which case the standard picture is complete and UFD's molecular extension has nothing to add, or it is not zero, in which case the residual is doing real work that the standard picture cannot account for.

Dr. S: And this is where the ASI and the RDI meet.

Dr. U: Yes. The ASI tracks the coherence of the nuclear foundation. Carbon-12 scores highly on that measure. Carbon-13 scores slightly lower. The RDI then tracks whether that nuclear difference shows up in the molecular bond.

Dr. S leaned forward slightly.
Now the theory was becoming dangerous in the right way.

Dr. S: Fine. Then state the actual prediction.

DR. U: After conventional reduced-mass corrections and the ordinary isotope effects have been removed, carbon-12 to carbon-12 bonds and carbon-13 to carbon-13 bonds should still show a small residual difference in their vibrational frequencies or bond energies.

DR. S: How small?

DR. U: On the order of a few tenths of a percent. Roughly 0.3 to 0.6 percent, depending on the type of bond and the molecular environment.

DR. S: That is specific enough to matter.

DR. U: It has to be. A vague claim that nuclear geometry might affect chemistry somehow would not be a real prediction. A specific number with a specific test condition is something experimenters can either find or fail to find.

That answer sharpened the whole conversation.
Up to that point, the chemistry bridge could still have been dismissed as a suggestive philosophical extension.
Once the carbon prediction appeared, the bridge became genuinely vulnerable.

Dr. S noticed that immediately.

DR. S: Let me be very clear. This is where your framework gets more interesting and more dangerous at the same time. The nuclear core can survive even if the chemistry bridge fails.

DR. U: Correct.

DR. S: Good.

DR. U: That distinction matters. Failure of the molecular extension would constrain or falsify the extension itself. It would not by itself kill the nuclear core. Each layer of the framework has to be tested on its own terms.

DR. S: Exactly. Because otherwise one bad extrapolation would make the whole framework look sloppier than it is.

DR. U: Right.

DR. S: But if the bridge succeeds—

DR. U: Then one geometric principle has crossed a very important threshold.

DR. S: From the nucleus into chemistry.

Dr. U: Yes.

That was the deepest point so far.

Standard chemistry was already predictive.

UFD did not deny that.

It claimed instead that predictive success had been achieved with an ontology too thin to reveal why the architecture of chemistry hangs together as it does.

If the carbon residual is real, that thinness becomes empirically demonstrable.

If it is not real, UFD has to retreat to the nuclear sector where its claims are stronger.

Dr. S let that sit and then moved to the broader physical consequences.

Dr. S: Fine. Let us suppose for the moment that the carbon residual is real. What else changes?

Dr. U: A great deal. First, bond stability becomes more physically layered. Not just electron bookkeeping, but electron structure plus nuclear congruence. Second, reaction pathways can no longer be thought of only as barrier heights on an abstract energy surface. They become changes in standing-wave organization constrained by inner geometry. Third, chemistry becomes a more natural bridge to materials science, because metallic bonding, delocalized resonance, and crystal-wide coherence are now all visible as continuations of the same underlying logic.

Dr. S: In other words, chemistry becomes less self-contained and more architectural.

Dr. U: Exactly. That is the gain if the bridge holds.

That was enough.

The conversation did not need to race ahead into proteins, life, or materials engineering.

It only needed to make one thing clear: if the atom is resonant and the bond is a real coherence merger, then the architecture of matter begins here.

Dr. S brought the exchange to its proper close.

Dr. S: Let me tell you what I think is strongest in your picture. You are not merely making chemistry more visual. You are trying to replace the bond as a formal bookkeeping relation with the bond as a physical coherence event.

Covalent, ionic, and metallic behavior become three different strategies for reaching lower strain — sharing, transfer, and delocalization. And the carbon case gives the extension a real empirical edge that ordinary interpretive preferences cannot deliver.

DR. U: Yes.

DR. S: That is a real gain if it works.

DR. U: Exactly.

DR. S: The danger is equally obvious. Standard chemistry is too successful to surrender anything cheaply. So your bond picture earns its place only if the residual effects are real, the diagnostics stay disciplined, and the bridge to materials science remains tighter rather than looser.

DR. U: Agreed.

DR. S: In other words: coherence language is not enough.

DR. U: Never enough.

That was the center of gravity.
Dr. U was not merely offering a more intuitive story about bonds.
He was claiming that one physical logic reaches from the nucleus, through the atom, into chemistry itself.
Dr. S was not merely defending orthodoxy.
He was insisting that the success of chemistry does not become more intelligible merely because a theory says it should.

Nothing about that was settled.
But the burden had become exact.

DR. S: Let me try the summary test.

DR. U: Go ahead.

DR. S: On your view, a chemical bond is a coherence merger between two resonant atoms. Its baseline stability still belongs to the electronic and orbital structure that standard quantum chemistry already handles well. But if the broader UFD extension is right, nuclear geometry contributes a dimensionless modifier to how fully that baseline stability is realized. Covalent bonds become shared standing-wave resonances, ionic bonds become transfer of resonant stability followed by source–sink attraction, and metallic bonds become crystal-wide delocalized resonance. RDI is a diagnostic of the extra coherence gain or frustration relative to baseline chemistry, not a standalone law. And the sharpest current test is carbon: after ordinary isotope effects are removed, carbon-12 to carbon-12 and carbon-13 to carbon-13 bonds should still differ by a small residual amount if nuclear congruence really reaches into chemistry.

DR. U: That is fair.

DR. S: The attraction is obvious.

DR. U: Yes.

DR. S: The danger is equally obvious.

DR. U: Yes.

DR. S: Good. Then the burden is finally clear.

INTERLUDE

By the end of the exchange, chemistry no longer looked like a layer of effective rules floating above a more basic atomic story.

At least in the UFD picture, the bond had become the place where the resonant atom turned outward and began to couple.

That was the gain.

The bond was no longer merely a bookkeeping success.
It was a physical merger of standing-wave systems.
Covalence, ionicity, and metallicity were no longer separate inventions of textbook classification, but three different ways that matter could seek lower-energy coherence.
And the sharpest empirical edge — the carbon isotope residual — made the chemistry bridge more than a suggestive metaphor.

That was also the risk.

Because the molecular extension is still just that: an extension. The nuclear core does not live or die with it. The chemistry bridge earns its place only if the residual effects are real, the diagnostics remain disciplined, and the added ontology continues to tighten rather than drift.

TEMPORARY VERDICT

Dr. U succeeded in making the bond physically legible. Chemical stability became a real coherence event rather than a merely formal one, and the three main bond strategies were gathered into one underlying logic.

Dr. S succeeded in holding the decisive line of pressure. A richer ontology of chemistry earns its place only if the molecular extension remains properly constrained, the carbon prediction survives scrutiny, and the bridge from nucleus to chemistry proves to be a genuine bridge rather than an attractive overreach.

The next burden now follows naturally.

IF BONDS CAN MERGE INTO LARGER COHERENT STRUCTURES, THEN THE NEXT QUESTION IS WHAT HAPPENS WHEN THAT LOGIC SCALES UP — INTO CONDUCTIVITY, MAGNETISM, SEMICONDUCTION, AND THE GEOMETRY OF MATERIAL PROPERTIES.

CHAPTER 16

The Coherence Spectrum of Matter

Conductivity, magnetism, and the possibility of control

Almost everything you interact with in ordinary life is condensed matter: solids and liquids rather than gases and plasmas. The phrase sounds technical, but it refers simply to matter in its dense, organized forms — metals, semiconductors, magnets, crystals, glasses, and the countless materials from which modern technology is built. Condensed matter physics is therefore not a niche subject. It is the physics of the everyday world, and it is one of the main reasons modern engineering is possible at all.

Its successes are immense. Band theory explained why some materials conduct electricity, others do not, and others sit between the two as semiconductors. Quantum models of spin and exchange helped explain magnetism. BCS theory explained superconductivity. Symmetry-breaking theory helped unify a wide range of phase transitions. The practical reach of these frameworks is extraordinary. Computers, chips, solar cells, batteries, medical imaging, and a great deal of modern manufacturing all rest on them.

The weakness of this success is not predictive failure. It is something quieter. The deeper physical picture often remains divided. Conductivity is taught through band theory. Magnetism is taught through exchange and spin alignment. Semiconductors are treated through threshold band structures. Superconductivity is treated through pairing and collective quantum order. These descriptions work very well, but they are often presented as distinct domains with partly overlapping mathematics rather than as visibly different expressions of one material principle.

UFD asks whether there is a more unified way to read them.

The previous chapters built the UFD picture of matter from the inside out. Atoms became coupled resonant systems. Bonds became coherence mergers. With that picture in place, the next question is what happens when not just two atoms, but enormous numbers of them, lock together in a regular arrangement. On the UFD view, a material is not just a pile of atoms held together by bonds. It is a lattice-scale coherence architecture built from resonant atoms and resonant bonds.

When many such bonds phase-lock across a crystal, the result is not merely a bigger molecule. It is a new level of organization. Collective standing-wave structures appear across the lattice, and those large-scale structures are what we observe as material properties. Conduction is one such regime. Insulation is another. Semiconduction is the threshold between them. Magnetism is another form of lattice-scale phase alignment. Superconductivity is the most dramatic case of all: a crystal-wide coherent state in which ordinary dissipation collapses. On this view, these are not separate miracles. They are different regimes of coherence.

Each familiar category then takes on a more specific physical meaning. A conductor is a material whose standing-wave structure extends across the crystal so that excitations can propagate from one side to the other. An insulator is a material whose resonances remain trapped and localized. A semiconductor lives near the threshold between those two states, where relatively small changes in conditions can shift the material from localized to extended modes. A magnet is a material whose local chiral or dipolar structures phase-lock over long distances. A superconductor is a crystal-wide coherence state in which the field flows without ordinary resistance.

None of this is meant to discard the mathematics of standard condensed matter theory. Band theory, exchange models, and superconducting formalisms still do their descriptive work. What changes is the ontology underneath them. Instead of a cluster of partly separate phenomena, matter begins to look like a spectrum of coherence states.

If that picture is right, it should do more than make matter easier to imagine. It should reveal new handles on matter itself — new ways of thinking about how to induce, suppress, or shift material states by acting directly on coherence. That is the burden of this chapter. Not merely whether the coherence-spectrum picture is elegant, but whether it begins to expose matter to forms of control that the more piecewise standard picture does not naturally foreground.

Maturity note. The coherence-spectrum picture reinterprets conductivity, insulation, semiconduction, magnetism, and superconductivity as different regimes of lattice-scale coherence. It is constrained by the framework, but it is not yet quantitatively closed at the level of independent material predictions. This chapter should therefore be read as a

By now the conversation had moved a long way from particles.
The atom had become a resonant system.
The bond had become a coherence merger.
So the next question was inevitable.

What happens when those mergers scale?

Dr. S began with the kind of skepticism that usually means a physicist is
about to defend a whole discipline at once.

DR. S: Fine. We have reached condensed matter, which means we have
reached one of the most successful parts of physics and engineering. So let
me ask the blunt question. What is a material on your view?

DR. U: A coherence architecture.

DR. S: Meaning?

DR. U: Not just a collection of atoms held together by bonds, but a lattice-
scale organization of resonant atoms and bonds. The properties we call
electrical, magnetic, mechanical, and superconducting are large-scale
expressions of how coherence propagates through that structure.

DR. S: So conductivity, insulation, semiconduction, magnetism, and
superconductivity are all supposed to be different phases of the same
underlying logic.

DR. U: Different regimes of the same logic, yes.

DR. S: Which is either elegant or dangerously overcompressed.

DR. U: As usual, yes.

That was the right place to begin.
The real claim was not that standard solid-state physics was wrong in its
equations.
It was that its successes might be the visible behavior of one deeper
coherence spectrum.

Dr. S pressed immediately on the most familiar property first.

Conductivity.

DR. S: All right. What is a conductor?

DR. U: A material whose valence standing waves overlap into a crystal-wide shared resonance.

DR. S: In plainer language.

DR. U: The outer standing-wave structure of each atom overlaps so well with its neighbors that the field can support one extended path across the entire material. One continuous channel through which excitations can travel from one end to the other. Standard theory calls that channel a conduction band. UFD says that band is a real sea of shared resonance.

DR. S: So a conductor is a material where the standing-wave system stops being local.

DR. U: Exactly. The crystal behaves like one connected field structure rather than a set of isolated atomic islands.

DR. S: And an insulator?

DR. U: The opposite regime. The resonances remain trapped and localized around individual atoms or small groups of atoms. There is no crystal-wide path available for stable transport.

DR. S: So the band gap becomes—

DR. U: A dead zone between trapped and extended resonance. The usual language survives. The ontology underneath it changes.

That immediately improved the conversation.
Conductors and insulators were no longer two separate textbook categories. They had become two ends of one coherence problem: whether resonance stays local or goes collective.

Dr. S recognized the gain and pressed where the ambiguity lived.

Threshold systems.

DR. S: Fine. Then semiconductors are the obvious in-between case.

DR. U: Exactly.

DR. S: And what changes in your telling?

DR. U: Standard theory already says semiconductors sit between localization and conduction. UFD keeps that, but makes the threshold more physical. A semiconductor is not merely a material with a certain band gap. It is a coherence-threshold system. It can be pushed toward trapped behavior or extended behavior because its standing-wave architecture is already close to tipping either way.

DR. S: So the ordinary distinction between conductor and semiconductor is not erased.

DR. U: No. It is deepened. The reason semiconductors matter so much technologically is that they already sit near a coherence boundary.

DR. S: Which means, on your view, they should be especially susceptible to new kinds of control.

DR. U: Yes. And that is one of the main burdens here. If matter really is a coherence spectrum, semiconductors should be one of the first places where that ontology cashes out experimentally.

That mattered.
Because the discussion had already begun to cross from interpretation into validation.
A new ontology earns almost nothing if it merely redraws known categories. It earns something when it suggests new handles on them.

Dr. S then turned to the next category.

Magnetism.

DR. S: All right. Then what is a magnet?

DR. U: A material in which many local rotational structures have phase-locked their order across long distances.

DR. S: So we are not starting from scratch here. We already treated magnetism as rotational flow in the field.

DR. U: Right. What matters here is how that local picture becomes bulk behavior. In most materials, local circulation patterns point in random directions and cancel out on average. In a magnet, they line up. Many small rotations point the same way. The result is a coherent macroscopic field with a stable direction.

DR. S: Standard theory would talk about exchange interaction and domain alignment.

DR. U: And both of those terms remain useful. UFD does not deny them. What it adds is the deeper question of what that exchange language is describing. On this view, it is the field's drive toward a lower-strain aligned rotational state.

DR. S: So exchange remains descriptively valid, but not fundamental.

DR. U: Exactly. The formalism survives. The ontology shifts.

For a moment the structure of the whole discussion became visible.
Conductor: extended resonance.
Insulator: trapped resonance.
Semiconductor: threshold resonance.
Magnet: aligned rotational resonance.
The theory was trying to replace a list with a spectrum.

Dr. S saw that clearly, which was why he moved to the place where the spectrum becomes most dramatic.

Superconductivity.

DR. S: Fine. Then let us get to the hard one. What is a superconductor?

DR. U: A crystal that has snapped into one large-scale coherent standing wave.

DR. S: That is a strong sentence.

DR. U: It should be.

DR. S: More specifically.

DR. U: In ordinary conduction, a material supports extended transport, but scattering and incoherence still cost energy. In superconductivity, something more organized happens. Below a critical temperature, the lattice and its mobile resonance structure enter a collective phase-coherent state. The material no longer carries many competing local modes that can scatter against each other. It sustains one larger coherent order. Resistance collapses because the usual dissipation channels are no longer available.

DR. S: Standard theory describes this through the BCS picture — Cooper pairs of electrons forming bound states that flow without scattering.

DR. U: Yes. And that picture remains descriptively powerful. But on the UFD view, the pairing is one signature of the state, not its deepest cause. The deeper claim is that the material has entered a crystal-wide coherence regime.

DR. S: And the Meissner effect?

DR. U: Natural on this view. A strongly coherent interior rejects incompatible magnetic disorder. Standard theory describes the effect through surface currents that cancel the internal field. UFD agrees those currents are real and calculable. What it adds is the deeper reason: the collective interior order will not sustain that internal disturbance, so the field is pushed out.

DR. S: So the superconductor is not merely a better conductor.

DR. U: No. It is a different coherence regime altogether.

That was one of the strongest moments in the exchange. Superconductivity had stopped being a weird exception in condensed matter.
It had become the clearest example of what the whole discussion was trying to say: material properties are coherence states.

Dr. S recognized that immediately. Which was why he sharpened the pressure where it mattered most.

DR. S: Let me state the obvious objection. All of this could still be relabeling. Band theory, exchange, pairing, lattice order — perhaps you are simply telling a more vivid story about results standard condensed matter theory already knows.

DR. U: That would be a fair objection if the discussion ended there.

DR. S: And does it?

DR. U: It should not. Because the point of an ontological shift is not only to clarify what is already known. It is to reveal new handles on matter that would otherwise remain hidden.

DR. S: Good. Now we are at the real issue.

That was the hinge.
If the theory did not imply new handles on matter, then the ontological gain would remain largely aesthetic.

Dr. S pressed exactly there.

DR. S: Fine. Then tell me what new controls become possible if your picture is right.

DR. U: Several, in principle. The first is field-controlled semiconduction. If semiconductors are threshold coherence systems, then one should not be limited to static doping and geometry alone. It should become possible to push a material across or away from a conduction threshold by properly tuned resonant fields.

DR. S: So not only static band engineering, but dynamic coherence control.

DR. U: Exactly.

DR. S: And superconductors?

DR. U: Similar logic. If superconductivity is a crystal-wide coherence state, then it should be actively inducible, stabilized, weakened, or quenched by the right resonant conditions. Not easily. But naturally, if the ontology is right.

DR. S: That is already more than reinterpretation.

DR. U: It has to be.

DR. S: What else?

DR. U: Coherence-guided material switching more broadly. Materials that today are treated as fixed categories may turn out to be more tunable than expected if coherence can be actively shaped rather than only passively accepted. The line between conductor and insulator, or between magnet and non-magnet, might become something that can be crossed in real time under the right field conditions.

That mattered because the theory had now crossed the line from "what matter is" to "what matter might be made to do."

Dr. S kept pressing. He should have.

DR. S: Fine. But now I want the caution stated clearly. This is where people start sounding like futurists rather than physicists.

DR. U: Then let us keep the stages clear.

DR. S: Good.

DR. U: Near-term: field-sensitive modulation of semiconductor thresholds and coherent transport regimes. Mid-term: better control of superconducting onset and suppression. Farther horizon: supermalleability and deliberately coherence-shaped materials.

DR. S: Supermalleability?

DR. U: A temporary reduction of coherence barriers that lets a material deform much more easily without ordinary fracture dominating. That is speculative, but it follows naturally from the ontology if the ontology is right.

DR. S: Go on.

DR. U: At the far horizon lies the idea of a fully coherence-optimized material — what in UFD is called Emetium.

DR. S: Hypothetical.

DR. U: Entirely. It is not a laboratory claim. It is the conceptual endpoint of the ontology — what would be possible if every coherence dimension were maximized in a single substance.

That improved the conversation.
A weaker version would have let the technology race ahead of the ontology. A stronger version kept the order strict: first explanation, then control, then staged exposure, and only finally the far horizon.

Dr. S turned back to the scientific burden.

DR. S: All right. Then what would actually count as validation here? Not in the broad poetic sense. In the hard sense.

DR. U: The ontology gains strength if new controls become possible specifically where the coherence picture says they should. If certain semiconducting materials proved more responsive to carefully tuned electromagnetic fields than the static band picture would predict, that would matter. If superconducting onset and suppression became more directly shapeable by resonant field conditions, that would matter. If certain materials could be moved across conductivity regimes through coherence shaping rather than only conventional thermal or chemical manipulation, that would matter.

DR. S: And if none of that happens?

DR. U: Then the ontology loses much of its practical force. It may still remain an interesting explanatory picture, but one of its strongest empirical promises weakens sharply.

DR. S: Good. Because that is the right kind of risk.

DR. U: It has to be the right kind. Otherwise, the whole coherence-spectrum idea remains too safe.

That was the center of gravity.
Dr. U was not merely offering a more unified language for condensed matter.
He was claiming that if the unity is real, then matter should become more controllable in ways that standard ontology does not naturally expect.
Dr. S was not merely defending standard theory.
He was insisting that a new ontology earns its place when it changes what can be done, not only what can be said.

Nothing about that was settled.
But the burden had become exact.

DR. S: Let me tell you what I think is strongest in your picture. You are replacing a cluster of material categories with one underlying spectrum. Conductors, insulators, semiconductors, magnets, and superconductors become different coherence regimes of one lattice-scale problem. That is a genuine gain in unity if it holds.

DR. U: Yes.

DR. S: And the strongest empirical stake is not the reinterpretation itself. It is the claim that new forms of control should follow from it.

DR. U: Exactly.

DR. S: The danger is obvious too. Condensed matter already has enormous practical success. So your ontology earns its place only if the control story eventually becomes more than a suggestive promise.

DR. U: Agreed.

DR. S: In other words: coherence language is not enough.

DR. U: Never enough.

That was the right note.
Not dismissal.
Not triumph.
Pressure.

So Dr. S gave the model its final test.

DR. S: Let me try the summary test.

DR. U: Go ahead.

DR. S: On your view, matter is not just a pile of atoms but a lattice-scale coherence architecture built from resonant atoms and bonds. Conductors are materials whose standing-wave structure extends across the crystal. Insulators are materials whose resonances remain trapped and localized. Semiconductors are threshold systems near the boundary between those two states. Magnets are materials whose local rotational or dipolar structures align coherently across the lattice. Superconductors are crystal-wide coherence states in which incoherent dissipation collapses. Standard condensed matter tools remain useful, but they are being re-read as descriptions of one deeper coherence spectrum rather than of several unrelated material miracles. And the strongest stake of the ontology is technological: if it is right, matter should become more controllable through field-shaped coherence than standard ontology alone would naturally suggest.

DR. U: That is fair.

DR. S: The attraction is obvious.

DR. U: Yes.

DR. S: The danger is equally obvious.

DR. U: Yes.

DR. S: Good. Then the burden is finally clear.

INTERLUDE

By the end of the exchange, matter no longer looked like a set of separate material categories stitched together by good mathematics and bad intuition. At least in the UFD picture, it had become a coherence spectrum. That was the gain.

Conductors, insulators, semiconductors, magnets, and superconductors were no longer fundamentally different kinds of matter. They were different large-scale regimes of how resonance organizes across a lattice. That gave the subject a new unity. It also linked the discussion directly to what came before: resonant atoms and bonds had now become the visible architecture of bulk material life.

But the more important gain was practical. If the ontology is real, then matter should not only become easier to describe. It should become easier to control. That was the sharp scientific stake. Field-controlled semiconduction, induced or suppressed superconductivity, and more general coherence-guided material switching are not decorative futures in this picture. They are the places where the ontology becomes empirically dangerous. If the controls do not appear, the coherence-spectrum story loses much of its force. If they do appear, then condensed matter may turn out to have been waiting not only for better equations, but for a better picture of what its equations are equations of.

TEMPORARY VERDICT

Dr. U succeeded in making matter physically legible. Bulk material properties became different regimes of one coherence spectrum rather than a cluster of unrelated special cases, and the theory's strongest promise emerged where it should: in the possibility of new control.

Dr. S succeeded in holding the decisive line of pressure. A unified ontology of matter earns its place not by elegance alone, but by whether its promised controls materialize in disciplined, staged, and experimentally meaningful ways.

The next question followed naturally.

IF MATTER REALLY IS COHERENCE STRUCTURED AT EVERY SCALE, THEN THE NEXT BURDEN IS LIFE: CAN METABOLISM, ENZYMES, AND BIOLOGICAL ORGANIZATION BE DESCRIBED IN THE SAME LANGUAGE WITHOUT DISSOLVING INTO HAND-WAVING?

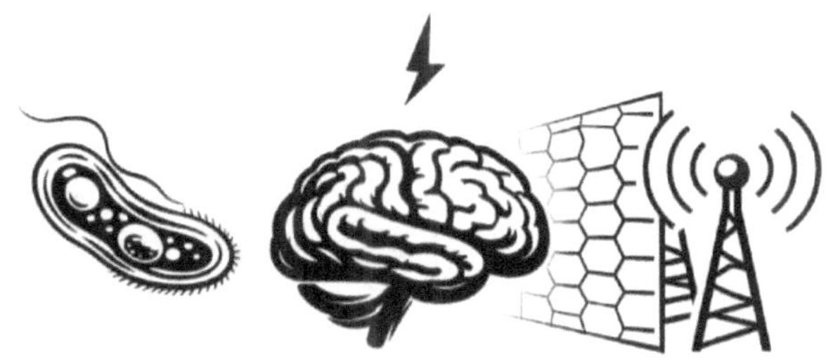

CHAPTER 17

Resonant Biochemistry

The cell, metabolism, and life as organized coherence

Life is the most organized thing we know of. A single cell contains millions of molecules, thousands of coordinated reactions, and a level of internal order so precise that the cell can grow, divide, repair itself, respond to its environment, and help reproduce its own form. And yet living systems are built from the same atoms, the same bonds, and the same chemistry as everything else. Whatever life is, it is something ordinary matter is somehow able to do under the right conditions.

The scientific discovery of life's molecular basis is one of the great achievements of modern science. Vitalism — the idea that life required a special non-physical principle — gradually collapsed as biochemistry and molecular biology advanced. Enzymes were isolated and studied. Metabolic cycles were mapped. DNA's structure was uncovered. The genetic code was deciphered. Membranes, signaling pathways, molecular motors, and protein folding were all brought into view. The result is a framework of extraordinary explanatory power. Modern medicine and biotechnology depend on it.

The weakness of that framework is not predictive failure. It is something quieter. A living cell can still seem, in the standard picture, like a brilliantly catalogued bag of chemicals. The reactions are known. The enzymes are known. The pathways are known. But what often remains less fully described is how all those local events hold together as one self-organizing whole. A cell is not just a list of reactions happening in a container. It is something that maintains its own organization against the constant tendency toward disorder.

That is where UFD enters. It does not seek to replace the chemistry of life. It asks whether the chemistry already belongs to a larger physical organization than the standard picture usually names.

The previous chapters built that question from the inside out. Atoms became coupled resonant systems. Bonds became merged resonances. Materials became coherence architectures. The next step is life. If matter is a coherence architecture, then the living cell — the most highly organized form of matter we ordinarily encounter — should also be a coherence architecture, not a different kind of physics but a more elaborate organization of the same physics.

On this view, the cell is not a passive soup of reacting parts. It is a resonant system. The membrane is not only a wall with selective channels, but a structured interface that helps regulate coherence between the inside and outside of the cell. The cytoskeleton is not only a scaffold, but also a possible pathway for organized vibrational and mechanical coordination. Enzymes are not only catalytic shapes, but highly selective structures that guide reactions through lower-disorder pathways. Metabolism is not crude burning of fuel, but the controlled release and transfer of stored organization. Signaling is not only message passing, but controlled reconfiguration of how the cell's parts are coupled.

Health, in this picture, is the preservation of multiscale organization. Disease is not only molecular malfunction, but also fragmentation or breakdown of that organization. UFD does not replace standard molecular accounts of illness. It adds a deeper layer in which life and health are also questions of coherence: how well the system maintains its resonant architecture across many interacting levels at once.

If that ontology is right, it should do more than make life easier to picture. It could suggest new ways of measuring coherence loss before it becomes overt dysfunction, and perhaps new ways of supporting biological order that work with the body's own organization rather than only through one molecular lever at a time. This is not yet a clinical program. It is an empirical direction the framework would have to develop if its account of life is to earn a place beside the molecular biochemistry that already explains so much.

By the time they reached life, the conversation had already passed through atoms, bonds, matter, and control. So the stakes had changed. This was not going to be a debate about whether chemistry was real. It was going to be a debate about whether chemistry is the whole ontology of life.

Dr. S began exactly where he should have.

DR. S: Fine. We have reached biochemistry, which means we have reached a field that already explains a great deal with ordinary chemistry. So tell me why I should want anything more.

DR. U: Because life is more organized than a parts list can explain by itself.

DR. S: That sounds like the beginning of a bad theory.

DR. U: It would be, if it denied the parts list. I am not denying enzymes, membranes, transport chains, receptors, metabolites, or pathways. I am saying the chemistry is real, but the chemistry is not the whole ontology.

DR. S: More specifically.

DR. U: The cell is not just a set of reactions happening in one place. It is a coherence-organized system. The question is not whether the molecules are there. It is what physically binds them into one stable, self-regulating whole.

That was the first important distinction.

Dr. U was not trying to replace biochemistry with mysticism.

He was trying to say what kind of physical system a cell would have to be if biochemistry were really going to hold together as one life process rather than as a long list of local events.

Dr. S pushed where that claim had to begin.

The cell itself.

DR. S: Fine. Then what is a cell?

DR. U: A resonant instrument.

DR. S: That is exactly the kind of sentence I do not trust unless you unpack it.

DR. U: Fair enough. The membrane is not merely a lipid envelope holding the cell's contents inside. It is a structured interface that regulates exchange, separation, and timing between inside and outside. The cytoskeleton is not just scaffolding. It is also a transport and coordination framework through which organized mechanical and vibrational patterns can spread. The cell's machinery is therefore not just floating in chemical soup. It is embedded in a coherence-shaped boundary system that helps give the whole cell physical unity.

DR. S: So the cell is not a bag of parts. It is a layered coherence device.

DR. U: Exactly.

DR. S: And that is meant literally, not poetically.

DR. U: Literally.

Dr. S did not push back immediately.

He knew that "the cell is an instrument" could either be empty rhetoric or the start of a real mechanism.

What mattered was whether the mechanism touched familiar biochemical work.

So he moved to the place where biochemistry first becomes operationally concrete.

Enzymes.

DR. S: All right. Then tell me what an enzyme is.

DR. U: A resonance filter.

DR. S: More specifically.

DR. U: An enzyme is a protein structure whose geometry lets it guide reactants through a lower-decoherence transformation pathway. Standard biochemistry says it lowers the activation barrier. UFD agrees, but asks why. On this view, the active site is not just a lock that fits a substrate. It is a highly selective cavity that grips, strains, aligns, and phase-guides the substrate through one preferred transformation.

DR. S: So catalysis becomes coherence guidance rather than merely barrier lowering.

DR. U: Exactly. Barrier lowering remains the standard description of what happens. Resonance guidance is the physical picture of how it happens.

DR. S: And you are not denying ordinary chemical kinetics.

DR. U: No. I am saying kinetics is tracking a deeper geometric filtering process.

That answer interested Dr. S more than it impressed him.
It was at least the right kind of move: not discarding the standard description, but trying to turn it into a physical mechanism.

So he pressed on the strongest test of that move.

DR. S: Fine. Then how would I know this is more than nicer language? "Enzymes as resonance filters" sounds good. What does it actually buy you?

DR. U: A control prediction.

DR. S: Go on.

DR. U: If an enzyme really depends on a specific resonant geometry, then external fields properly matched to that geometry should be able to inhibit or enhance its function without acting as conventional chemical inhibitors. The standard way to inhibit an enzyme is to add a molecule that blocks its active site or alters its shape. UFD predicts that there should be another way: carefully tuned electromagnetic fields, matched to the enzyme's resonant signature, should be able to interfere with or reinforce its operation directly.

DR. S: You mean field-based catalysis control.

DR. U: Exactly. Destructive interference should jam the active site. Constructive interference should reinforce it.

DR. S: That is a real prediction.

DR. U: It has to be.

DR. S: And this is what you call the Resonant Modulation Hypothesis.

DR. U: Yes. Not a completed platform, but a real falsifiable direction. If carefully tuned fields cannot measurably modulate enzyme behavior under good controls, then one of the model's sharpest claims weakens immediately.

A less serious approach would have stopped at philosophical reinterpretation.
A stronger one turned the reinterpretation into a falsifiable control claim.

Dr. S respected that enough to keep pressing seriously.
He moved next to metabolism.

DR. S: All right. Then what is metabolism in your language?

DR. U: A resonant energy circuit.

DR. S: That is another phrase I am going to need translated.

DR. U: Standard metabolism looks like a sequence of reactions that extracts energy from molecules. UFD says the deeper physical picture is the controlled release of stored geometric coherence. Fuel molecules are not just arbitrary collections of bonds. They are stable structures whose internal organization stores usable order. Metabolism is the stepwise liberation of that order into work.

DR. S: So the cell is not "burning fuel."

DR. U: Not in the crude sense. It is disassembling coherent structures in a controlled way, extracting their organization piece by piece rather than releasing it all at once.

DR. S: Controlled demolition.

DR. U: Exactly.

Dr. S nodded once.
The term was vivid, but not empty.
He could see the appeal: the model was trying to explain why metabolism unfolds through constrained stages rather than as one chaotic release of energy.

So he pushed into the first iconic example.

Glucose.

DR. S: Fine. Then what is special about glucose?

DR. U: It is a good coherence-storage molecule.

DR. S: More specifically.

DR. U: Its ring geometry is stable, symmetric, water-compatible, and dynamically reversible. That makes it an effective molecular reservoir: stable enough to store energy, flexible enough to release that energy through controlled transformation, and legible enough for enzymes to recognize and process.

DR. S: So glucose is not merely a fuel chosen by evolutionary happenstance.

DR. U: Not on this view. It is a geometry well suited to resonant storage and staged release.

DR. S: And glycolysis is the staged release.

DR. U: Yes. Glycolysis breaks glucose down through a tightly ordered sequence, and UFD treats each enzyme in that sequence as a resonant catalyst tuned to one specific transformation. The pathway breaks the larger coherent structure into smaller fragments while releasing stored energy in controlled increments.

That answer changed the tenor of the conversation.
Metabolism was no longer just being redescribed as "energetic coherence."
Specific molecules and pathways were being drawn into the picture.

Dr. S pushed further inward.

Mitochondria.

DR. S: All right. Then what is a mitochondrion?

DR. U: A toroidal engine of metabolic coherence.

DR. S: That is a very UFD sentence.

DR. U: It is also the right one. Standard biochemistry calls the mitochondrion the powerhouse of the cell, and that is true as far as it goes. UFD agrees, but adds that the folded architecture matters physically. The cristae are not just there to pack more chemistry into less space. They create a geometry that stabilizes and cycles energetic flow in a particular way.

DR. S: So the shape is not incidental to the function.

DR. U: Exactly. Form dictates function much more literally here than the standard summary language often suggests.

DR. S: And ATP synthase?

DR. U: One of the clearest examples of resonant machinery in biology. It literally rotates as protons flow through it. Standard biochemistry already describes that beautifully. UFD adds that this is one of the clearest cases of organized field flow being converted directly into usable biochemical work.

DR. S: So the electron transport chain becomes—

DR. U: A field-guided resonator. The standard molecular biology stays intact. What UFD adds is the claim that the entire chain functions as a coherent system, not merely as a series of independent reactions that happen to be linked.

That answer was more concrete than many biological reinterpretations manage to be.
It did not say "life is special" and stop there.
It said: if geometry matters, then organelles should look like machines whose form is causally significant.

Dr. S noticed that and shifted to the broader systems problem.

Signal transduction.

DR. S: Fine. Let us move from metabolism to signaling. Standard biology says cells communicate through receptors, ion channels, phosphorylation cascades, and second messengers. What does your picture add?

DR. U: It adds the idea that signaling is not just molecular relay. It is phase organization across a boundary system — a coordinated reorganization of the cell's internal rhythms in response to information from outside.

DR. S: More specifically.

DR. U: Receptors are still receptors, ion channels are still ion channels, phosphorylation cascades are still real, and second messengers still carry information. UFD does not deny any of that. But the deeper point is that cells must preserve coherence while changing state. A cell receiving a signal does not just trigger a chain of reactions. It reorganizes the way its resonant components are coupled to one another. Signaling is therefore not merely message passing. It is controlled reconfiguration of a coherence network.

DR. S: So the molecular steps remain, but they are nested inside a larger dynamical picture.

DR. U: Exactly. The chemistry is real. The chemistry is not the whole ontology.

That was the model's first true systems-level move.
Enzymes, metabolism, and signaling were no longer three separate topics. They were becoming different expressions of one claim: life is not just chemistry, but chemistry under active coherence management.

Dr. S heard that clearly and pushed where such a claim becomes dangerous. Disease.

DR. S: All right. Then what is pathology?

DR. U: A breakdown in coherence.

DR. S: That is dangerously broad.

DR. U: It needs to be said carefully. I am not saying every disease can be reduced to one slogan. I am saying many biological failures can be re-read physically as failures to maintain coherence across multiple scales at once. Membranes can lose organized gradients. Enzymes can lose the precise alignment that lets them function selectively. Mitochondria can lose the structural coherence that makes energy production efficient. Tissues can lose the coordinated rhythms that let their cells act as one functional unit.

DR. S: So inflammation, metabolic dysfunction, maybe cancer—

DR. U: Yes. Not as "nothing but" coherence failure, but as states in which coherence failure becomes physically central to what goes wrong. The molecular descriptions remain valid. The mutations are real. The signaling errors are real. UFD is not denying any of that. It is saying there is often a deeper layer of organizational breakdown that the molecular catalogue alone does not name.

DR. S: And your claim is that ordinary biochemical imbalance is then one visible expression of a deeper structural disorder.

DR. U: Exactly.

This was a dangerous turn, and the model handled it correctly only by narrowing the claim.
Pathology was not being reduced to one slogan.
The proposal was that many disease states may involve coherence failure as a physically important layer of what goes wrong.

Dr. S then pushed where the whole program had to become experimentally vulnerable.

DR. S: Fine. Then how does any of this become more than a suggestive philosophy of life?

DR. U: Through control and measurement.

DR. S: More specifically.

DR. U: If life depends partly on coherence organization, then biological systems should show measurable coherence lifetimes, not just chemical concentrations. There should be ways to observe how long a structure maintains its coherent state, and those lifetimes should correlate with biological function. Disease states should correlate with shorter coherence persistence in the relevant systems. And interventions should eventually be judged not only by whether they change chemistry, but by whether they restore coherent organization.

DR. S: So diagnosis becomes coherence mapping.

DR. U: In the stronger program, yes.

DR. S: And therapy becomes coherence modulation.

DR. U: Exactly — where possible, and where the system really does depend on resonance organization.

That was one of the most important transitions in the whole exchange.
Biochemistry was no longer only an interpretive domain.
It had become an engineering and diagnostic burden.

Dr. S noticed that and, predictably, pushed on the discipline of the claim.

DR. S: Let me be careful here. This is where a lot of grand biological theories lose me. Once people say "life is coherence," they often start promising universal healing technologies and magical field medicine five minutes later. Are you making those claims?

DR. U: Not as mature results, no.

DR. S: Good.

DR. U: The right claim is more modest and more serious. If the ontology is right, then coherence-sensitive imaging, resonance-based enzyme modulation, and field-guided biomaterials become experimentally meaningful directions. But these are directions, not completed platforms. None of them is a clinical reality today. None of them should be promised as one.

DR. S: Better.

DR. U: The biology is too complex for triumphalism. The only honest path is tiered: first interpretation, then controlled diagnostics, then limited intervention, then maybe broader engineering later.

DR. S: That is the first fully sane sentence anyone says in these domains.

DR. U: It has to be sane or it is not worth saying.

That answer made the model much stronger.
A sloppier framework would have overclaimed.
A more serious one distinguished clearly between interpretation, diagnostic consequence, and technological horizon.

Dr. S then widened the frame one final time.

DR. S: Let me tell you what I think is strongest here. You are not denying molecules, enzymes, receptors, organelles, pathways, or signaling networks. You are trying to say what physically binds them into one organized living process.

DR. U: Yes.

DR. S: That is much stronger than saying "the cell has rhythms."

DR. U: It also carries a stronger burden. If coherence is truly the relevant physical variable, then it must become measurable, manipulable, and tightly linked to both function and pathology.

DR. S: So biochemistry becomes a bridge domain.

DR. U: Exactly.

DR. S: Between chemistry and life.

DR. U: Yes.

That was the center of the exchange.
Not whether cells have oscillations.
Everyone already knew that.
The question was whether oscillation, phase, and coherence are part of the physical infrastructure of life itself.

Dr. S still trusted standard biochemistry more than the replacement ontology.
Dr. U still thought the replacement ontology gave the deeper picture.

Nothing about that was settled.
But the burden had become exact.

DR. S: Let me try the summary test.

DR. U: Go ahead.

DR. S: On your view, the cell is not a bag of chemicals but a resonant instrument. The membrane is a resonant interface, the cytoskeleton a vibrational and transport guide, and the molecular machinery of the cell is embedded in a coherence-organized boundary system. Enzymes are resonance filters that guide substrates through lower-decoherence pathways. Metabolism is controlled release of stored geometric coherence rather than crude burning. Glucose is a good coherence-storage molecule, glycolysis is a staged resonant disassembly, and the mitochondrion is a toroidal engine of metabolic coherence. Signal transduction is not merely relay chemistry, but controlled reconfiguration of a coherence network. Pathology is not reducible to one slogan, but many disease states can be re-read as multiscale failures of coherence maintenance. And the strongest empirical burden is that this should lead to measurable coherence lifetimes, resonance-sensitive diagnostics, and bounded field-based modulation of biological function.

DR. U: That is fair.

DR. S: The attraction is obvious.

DR. U: Yes.

DR. S: The danger is equally obvious.

DR. U: Yes.

DR. S: Good. Then the burden is finally clear.

By the end of the exchange, life no longer looked like a fortunate heap of successful chemistry.

At least in the UFD picture, it had become chemistry under active coherence management.

That was the gain.

The cell was no longer merely a container.
Enzymes were no longer only rate accelerators.
Metabolism was no longer only energy extraction.
Signal transduction was no longer only message relay.
And pathology was no longer only a local chemical imbalance.
All of them had been drawn into one picture of organized field-supported function.

That was also the risk.

A richer ontology of life earns nothing by sounding holistic. It earns its place only if coherence becomes a real control variable: something measurable in healthy and diseased systems, something that can be mapped, and something that can sometimes be modulated under good controls. The Resonant Modulation Hypothesis mattered for exactly that reason. It turned the framework from a philosophical extension into an experimentally exposed research program.

TEMPORARY VERDICT

Dr. U succeeded in making biochemistry physically legible. The cell became a coherence-organized system rather than a passive mixture, enzymes became resonance filters, metabolism became controlled release of stored coherence, and pathology became a multiscale failure of organization rather than only a list of local faults.

Dr. S succeeded in holding the decisive line of pressure. A deeper ontology of life earns its place only if it keeps the chemistry, avoids vitalistic overreach, and turns coherence into a measurable and manipulable variable rather than a decorative word.

The next question now followed naturally.

IF LIFE DEPENDS ON COHERENCE ORGANIZATION ACROSS SCALES, THEN THE NEXT BURDEN IS THE BRAIN: CAN NEURAL ACTIVITY,

PATHOLOGY, AND COGNITION BE DESCRIBED IN THE SAME
LANGUAGE WITHOUT COLLAPSING EITHER INTO CHEMISTRY ALONE
OR INTO MYSTICISM?

CHAPTER 18

Neuroscience

Brain coherence and resonant neurophysics

The brain is the most complex object we know of. It contains tens of billions of neurons, each linked to thousands of others, forming a network of staggering density. From this organized activity come perception, memory, emotion, language, attention, decision, and the felt experience of being a self in a world. Few things in science are more familiar in consequence and more difficult in mechanism. We know an enormous amount about the brain, and yet the basic fact that matter can become aware of anything at all still feels astonishing.

Modern neuroscience earned its place through a long sequence of real discoveries. Neurons were identified as discrete cells. Synapses were shown to be sites of communication. Action potentials, neurotransmitters, plasticity, and large-scale brain imaging gradually turned the brain from a philosophical mystery into a scientific object. The framework that emerged is powerful: neurons fire, synapses strengthen or weaken, networks process information, and distributed regions coordinate behavior. Brain rhythms are well established too, but they are often treated as signatures of underlying computation rather than as part of the primary physical scaffolding of mind.

That framework is not weak. It is one of the great achievements of modern science. The question here is not whether neurons, synapses, neurotransmitters, and networks are real. They are. The question is whether they already belong to a deeper physical organization than the standard language usually emphasizes.

UFD keeps the machinery and changes the physical picture underneath it.

The previous chapters built that move from the inside out. Atoms became coupled resonant systems. Bonds became merged resonances. Materials became coherence architectures. Living cells became resonant instruments. The next step is the brain. If life is organized matter, then the brain is the place where that organization reaches its highest known density. UFD therefore asks whether the brain should be understood not only as an electrochemical organ, but as a nested coherence system.

On this view, neurons and neurotransmitters still do everything standard neuroscience says they do. Action potentials still propagate. Synapses still strengthen and weaken. Networks still process information. But what those structures are doing, at a deeper level, is organizing oscillatory field structure across multiple scales. The firing of individual neurons is not denied. It is re-situated inside a larger architecture of rhythm, phase, and coordinated oscillation.

Brain rhythms are therefore not background music while the real work happens elsewhere. They are part of the infrastructure of cognition itself. Alpha, beta, gamma, theta, and delta rhythms are not merely markers of brain state. On the UFD view, they help constitute the organized states they are usually taken to report. When distant regions oscillate in phase, they become functionally bound in a way that wiring alone does not fully capture. When their phases drift apart, they can become functionally disconnected even if the anatomical connections remain.

Mental clarity, memory, attention, and unified experience then become questions of multi-scale alignment. A focused mind is not only a brain with the right neurons active. It is a brain whose oscillatory layers are coordinating effectively across scales. Pathology, correspondingly, is not only chemical deficit or circuit damage. UFD adds a deeper layer in which many disorders can be understood as failures of coherence maintenance: fragmentation, rigidity, or runaway synchrony in the brain's oscillatory architecture. This does not replace the molecular and circuit-level descriptions. It adds another level on which those failures can be understood.

In this chapter, the discussion remains at the level of brain organization. The lighter field that has carried the explanatory work throughout the book is treated as the field organizing neural coherence here as well. The deeper coordination field is held back until the explicit discussion of consciousness and agency in Chapter 19. That distinction matters. It keeps the neuroscience argument from claiming more than it has yet earned.

Maturity note. Resonant Neuroscience keeps the established machinery of neuroscience and reinterprets the brain as a nested coherence system. Brain oscillations are well established empirically; what UFD adds is a claim about their physical status as infrastructure rather than epiphenomenal signatures. The stronger claims in this chapter

By the time the conversation reached the brain, the framework had built enough cumulative structure that the next move was natural rather than surprising. The question Dr. S and Dr. U were about to confront was not whether the brain has oscillations — everyone already knows it does — but whether those oscillations are part of the physical infrastructure of mind itself, or merely signs of work happening somewhere else.

Dr. S began with unusual caution.

DR. S: All right. We have reached the brain, which means we have reached the place where bad theories go to multiply. Neuroscience already has real successes. It knows about cortical maps, Hebbian learning, oscillations, neuromodulators, large-scale networks, pathology. So why should anyone want a deeper ontology here?

DR. U: For the same reason as before. Because a parts list is not yet a physical picture of integration.

DR. S: That is a sentence I am prepared to distrust.

DR. U: Fair enough. Then more sharply: standard neuroscience is very good at telling you which structures matter and which signals correlate with which functions. It is much less good at telling you what physically binds a moment of thought into one coordinated act across scales. You can map every region, record every spike, identify every neurotransmitter, and still not say what makes the result one moment of experience rather than a billion separate events that happen to share a skull.

DR. S: So your complaint is not that the circuitry is fake.

DR. U: Not at all. The circuitry is real. The question is whether circuitry alone is the deepest physical description of what the brain is doing.

That was the chapter's first crucial distinction.
UFD was not denying the machinery of neuroscience.
It was asking whether the machinery itself is organized by a deeper physical variable.

Dr. S did not deny that the brain is integrated.
Dr. U did not deny that standard neuroscience has earned its machinery.
The disagreement was whether integration itself deserved a stronger physical story than "many signals coordinated at once."

So Dr. S went straight to the basic object.

DR. S: Fine. Then what is a brain, on your view?

DR. U: A resonant organ.

DR. S: More specifically.

DR. U: A nested system of oscillatory structures. Neurons are not just wires that spike. They are local resonance units. Brain regions are not just modules with specialized jobs. They are larger attractors or cavities of coordinated oscillation. What we call cognition is the alignment and controlled misalignment of those oscillations across scales.

DR. S: So the brain is less a computer than an orchestra.

DR. U: In the right sense, yes. Not because it lacks mechanism, but because the mechanism is rhythmic and field-organized rather than purely symbolic. A computer processes information by passing discrete signals through circuits. An orchestra produces music by coordinating many instruments through shared timing, rhythm, and phase. The brain, on this view, is closer to the orchestra. The signals are real, but the coordinated rhythmic structure is what cognition actually consists of.

DR. S: And the electrochemistry?

DR. U: Still essential. But it is the instrument-making layer, not the whole music. You cannot have an orchestra without instruments, and you cannot have cognition without the electrochemical machinery of neurons. But the instruments alone do not make music.

That answer made the model legible immediately.
The claim was not that the brain stops being physical.
It was that the brain's physics is more oscillatory and field-organized than standard explanations usually emphasize.

Dr. S pressed where the idea had to become concrete.

The neuron.

DR. S: Fine. Then what is a neuron?

DR. U: A resonance unit.

DR. S: Meaning?

DR. U: A neuron does not merely carry a signal. It has an internal resonant profile, and that profile matters for what it can amplify, suppress, or phase-lock with. The membrane still matters. Ion channels still matter. Synapses still matter. But beneath those familiar mechanisms is a vibrational structure that helps determine how the neuron participates in larger coherence patterns.

DR. S: And what sets that profile?

DR. U: Several layers. The membrane and synapse shape the immediate electrical response. Neurotransmitters modulate the intercellular field relations. And more internally, microtubules help complete the neuron as an oscillatory unit.

DR. S: Microtubules.

DR. U: Yes. But not as mystical quantum strings doing all the thinking. Some theories try to load the whole problem of consciousness onto microtubules. UFD does not. In this picture, microtubules are intracellular resonance conduits. They help give the neuron a stable internal vibrational architecture. They are not the whole theory of mind.

DR. S: So you are not making microtubules the seat of consciousness.

DR. U: No. They have a real role, but not the whole burden.

This was an important act of restraint.
Microtubules were being given a role, but not the entire explanation.
That makes the model much stronger than views that try to explain everything with one subcellular structure.

Dr. S then pushed to the next obvious layer.

Neurotransmitters.

DR. S: All right. Then what are neurotransmitters doing in this picture besides what neuroscience already says they do?

DR. U: They are modulators of resonance.

DR. S: More specifically.

DR. U: Standard neuroscience already knows that neurotransmitters tune neural activity. UFD keeps that. But it adds a more physical picture. Different neurotransmitters bias the local circuit toward different coherence states. Excitatory systems amplify phase alignment. Inhibitory systems buffer, delay, or sculpt it.

DR. S: So glutamate and GABA stop being only "go" and "stop" chemicals.

DR. U: Exactly. Glutamate promotes excitatory coherence. GABA and glycine provide coherence buffering and phase shaping. The point is not merely that they alter firing probability. They help determine the oscillatory regime the circuit can sustain.

DR. S: So inhibition is not just suppression.

DR. U: No. It is part of how the system avoids destructive synchrony and carves useful rhythms out of raw excitation. A brain that was nothing but excitation would collapse into seizure-like states almost immediately. Inhibitory systems are part of what keep activity organized rather than chaotic or frozen.

That was one of the model's stronger moves.
The gain here was not that neurotransmitters became mysterious field entities.
It was that familiar excitatory and inhibitory chemistry was being re-read as modulation of coherence regimes rather than only of firing probability.

Dr. S respected that enough to push to the larger-scale architecture.

The brain itself.

DR. S: Fine. Then tell me what the major brain structures are doing, if not just carrying out specialized functions in the ordinary way.

DR. U: They are doing that too. But they are also acting as resonant subsystems with different roles in the overall coherence architecture.

DR. S: Give me examples.

DR. U: The hypothalamus acts as a master oscillator of bodily rhythm and motivational state. The hippocampus helps stabilize memory and contextual resonance. The thalamus is central in routing and synchronizing large-scale cortical rhythms. The cerebellum refines timing and predictive coordination. The brainstem ties higher structures to autonomic rhythms.

DR. S: So the anatomy stays. The physics underneath it thickens.

DR. U: Exactly. The brain is not reduced to one oscillator. It is a hierarchy of coupled oscillators with specialized geometries and timescales.

That was important.
The model did not have to abolish localization in order to deepen it.
It only had to say that localization alone was not enough.

Dr. S then moved to where modern neuroscience itself has become most comfortable talking in rhythmic terms.

Large-scale networks.

Standard neuroscience already speaks comfortably about large-scale networks. The default mode network, executive network, and salience network are now familiar parts of the field's language. UFD's stronger claim is that these are not just statistical clusters in imaging data. They are physically meaningful resonant modes of the system.

DR. S: All right. Then let us talk about those networks. What changes in your picture?

DR. U: They stop being only statistical clusters and become real large-scale resonant modes.

DR. S: Meaning?

DR. U: The default mode network is not just "the introspection network." It is a slower coherent mode suited to internal narrative, memory retrieval, and self-modeling. The executive system is a faster, more externally oriented coherence mode. And the salience network acts as a switching system that helps move the brain between them.

DR. S: So transitions between networks become—

DR. U: Phase shifts in a large-scale standing-wave architecture.

DR. S: That is one of your cleaner moves.

DR. U: It should be. Standard neuroscience already points toward rhythms and phase structure here. UFD is simply saying those are not secondary decorations. They are part of the mechanism.

For a moment the conversation relaxed.
Not because the disagreement weakened, but because the theory was now saying something intelligible about familiar mental functions without yet leaping into mysticism.

Before the conversation could go further, the idea of phase needed to be made plain.

Phase is a property of waves and oscillations. It refers to where a wave is in its cycle at a given moment — at the peak, at the trough, or somewhere in between. When two waves are in phase, they reinforce one another. When they are out of phase, they cancel. Phase-locking means two oscillating systems maintain a stable timing relationship. UFD's claim is that this phase coordination is not a side effect of brain activity. It is part of the infrastructure that makes coherent thought possible.

Dr. S then pushed into pathology, where a coherence-based ontology had to become more than beautiful.

DR. S: Fine. Then what is mental disorder on this picture?

DR. U: Broken resonance.

DR. S: That is dangerously broad.

DR. U: It needs to be said carefully. I am not saying every disorder can be collapsed into one slogan. I am saying many neurological and psychiatric failures can be re-read as failures in coherence architecture. The chemistry matters. The circuitry matters. But what often goes wrong physically is the stability of phase organization across scales.

DR. S: Give me examples.

DR. U: Attention disorders become unstable gating and wandering coherence. Seizures become pathological over-synchronization. Dissociation becomes loss of large-scale phase integration. Neurodegenerative decline becomes collapse of coherence persistence in key attractor systems.

DR. S: So too little coherence and too much coherence can both be pathological.

DR. U: Exactly. The issue is not "more coherence good." It is the right coherence, at the right scale, with the right flexibility. A healthy brain is not maximally synchronized. It is metastable — coherent enough to bind activity together when needed, flexible enough to shift between patterns when conditions change.

That answer made the conversation stronger.
A less disciplined framework would have equated health with generic harmony.
A stronger one recognized that living systems need flexible, metastable coherence rather than simple uniform synchrony.

Dr. S noticed that and pressed exactly where the framework needed pressure.
ADHD and wandering attention.

DR. S: All right. Then let us take a concrete case. Attention instability. What would you say about something like ADHD in this framework?

DR. U: A failure of coherence retention and gating.

DR. S: Meaning?

DR. U: The system cannot stably maintain one attractor — one preferred rhythmic pattern — long enough without slipping into competing oscillatory pathways. The problem is not simply insufficient effort or too little neurotransmitter in the crude sense. It is unstable phase selection. The brain keeps wandering because the relevant coherence basin is too shallow or too easily disrupted.

DR. S: So the pathology is real, but the interpretation changes.

DR. U: Exactly. Chemistry still matters because chemistry affects coherence. Stimulants may work in part because they deepen the relevant attentional basins. But the thing going wrong physically is not exhausted by chemistry alone.

That was a useful moment because it showed what the framework was trying to do in neuroscience more generally.
Not abolish standard mechanisms.
Reorganize them around a deeper physical variable.

Dr. S then moved to the most practical question.
Testing.

DR. S: Fine. How does any of this become more than suggestive language? "Brain coherence" is easy to say. What would count as real evidence for your picture?

DR. U: Three things. Better mapping, better diagnosis, and better control.

DR. S: More specifically.

DR. U: First, if cognition is phase architecture, then meaningful mental states should be distinguishable by global coherence geometry, not just by regional activation maps. Second, pathology should correlate with measurable disruption in coherence lifetimes and cross-frequency structure. Third, bounded resonance-tuned interventions should be able to nudge systems back toward healthier attractor organization.

DR. S: So not just "EEG shows oscillations." Stronger than that.

DR. U: Much stronger. The claim is that cognition and pathology should become legible as coherence topologies, and that treatment should gradually become more phase-sensitive and attractor-sensitive.

That was the model's first true engineering edge.
Neuroscience was no longer only interpretive.
It had become a claim about what neurodiagnostics and neuromodulation ought to become if the ontology were right.

Dr. S, predictably, pushed on the danger.

DR. S: Let me be careful here. This is where lots of brain theories become irresponsible. The moment someone says "coherence," they start promising miracle interventions, thought reading, or universal cures. Are you making those claims?

DR. U: Not as mature results, no.

DR. S: Good.

DR. U: The right claim is more modest and more serious. If the ontology is right, then field-sensitive diagnostics and bounded neuromodulation should become more physically intelligible and more targeted. But the brain is too complex for triumphalism. The path has to be staged: interpretation first, better mapping second, narrow control experiments third, and only later broader technologies if the early layers survive.

DR. S: Better.

DR. U: It has to be better. Otherwise, this becomes just another grand theory of mind that never exposes itself.

That answer improved the credibility of the model significantly.
A sloppier framework would have overclaimed.
A more serious one distinguished clearly between interpretation, diagnostic consequence, and technological horizon.

Dr. S then turned to the broader philosophical comparison.

DR. S: Let me tell you what I think is strongest here. You are not denying neurons, synapses, transmitters, or networks. You are trying to say what physically binds them into unified episodes of mind.

DR. U: Yes.

DR. S: That is much stronger than saying "the brain has rhythms."

DR. U: It also carries a stronger burden. If coherence is truly the relevant physical variable, then it must become measurable, manipulable, and tightly linked to both cognitive states and pathology.

DR. S: Which means neuroscience becomes a bridge domain.

DR. U: Exactly.

DR. S: Between chemistry and consciousness.

DR. U: Yes.

That was the sharpest point of the exchange.
Not whether the brain has oscillations.
Everyone already knew that.
The question was whether oscillation and phase were merely accompaniments to cognition, or the physical infrastructure of cognition itself.

Dr. S still trusted standard neuroscience more than the replacement ontology.
Dr. U still thought the replacement ontology gave the deeper picture.

Nothing about that was settled.
But the burden had become exact.

DR. S: Let me try the summary test.

DR. U: Go ahead.

DR. S: On your view, the brain is a nested coherence system rather than just a biochemical machine. Neurons are local resonance units, brain regions are larger attractors, and rhythms are part of the physical scaffolding of cognition rather than secondary signatures of it. Neurotransmitters modulate coherence rather than merely carrying messages. Large-scale networks are real resonant modes. Pathology is often best understood as disrupted, rigid, or runaway phase organization across scales. And the strongest empirical burden is that cognition and disorder should become more legible as coherence topologies, with diagnosis and intervention gradually becoming more phase-sensitive and attractor-sensitive if the ontology is right.

DR. U: That is fair.

DR. S: The attraction is obvious.

DR. U: Yes.

DR. S: The danger is equally obvious.

DR. U: Yes.

DR. S: Good. Then the burden is finally clear.

INTERLUDE

By the end of the exchange, the brain no longer looked like a machine whose rhythms were merely decorative side effects.

At least in the UFD picture, it had become a nested coherence organ.

That was the gain.

Neurons were no longer only spiking units.
Neurotransmitters were no longer only chemical messengers.
Networks were no longer only statistical groupings.
And pathology was no longer only a defect in chemistry or circuitry.
All of them had been drawn into one picture of brain function as organized oscillatory architecture.

That was also the risk.

A coherence-based neuroscience earns nothing merely by sounding more integrated or more satisfying. It earns its place only if coherence becomes a real physical variable: something that can be measured, tracked across brain states, compared across healthy and disordered systems, and cautiously influenced under disciplined conditions. If that never happens, the ontology remains interesting but insufficiently exposed. If it does, then neuroscience may prove to have been waiting not only for better maps of the brain, but for a deeper account of what those maps are actually tracing.

TEMPORARY VERDICT

Dr. U succeeded in making the brain physically legible. Neural activity, cognition, and pathology were drawn into one field-organized picture without discarding the successful machinery of neuroscience.

Dr. S succeeded in holding the decisive line of pressure. A deeper ontology of the brain earns its place only if it becomes experimentally disciplined rather than merely more satisfying. The next question now pressed in from just beyond the edge of the conversation.

245

CHAPTER 19

Consciousness

Proper Time and The Hard Problem of Consciousness

By the time the argument reaches consciousness, the burden changes. Up to this point, the book has been asking whether matter, life, and the brain can be made more physically intelligible by treating them as coherence architectures rather than as collections of disconnected parts. But consciousness introduces a different kind of question. The problem is no longer only how a system coordinates itself. It is how there comes to be a point of view at all — a center to whom the coordination is present as experience.

Modern neuroscience explains a great deal about the brain. It maps regions, tracks signals, identifies neurotransmitters, describes plasticity, and increasingly reveals the oscillatory structure of large-scale brain activity. That achievement is real and indispensable. But even a very detailed neuroscience still leaves a familiar gap. It can describe how the brain processes information, integrates inputs, guides action, stores memory, and stabilizes attention. It is less clear on how any of that processing becomes a lived perspective — why there is something it is like to be the system rather than only a highly organized mechanism doing what it does.

That gap is what philosophers of mind have come to call the hard problem of consciousness. The so-called "easy" problems are not easy in practice, but they are at least clear in form: how the brain discriminates stimuli, integrates information, reports its states, directs behavior, or binds sensory inputs into coordinated activity. Those are problems about function, mechanism, and organization. The hard problem is different. It asks why any of those functions should be accompanied by experience at all. Why should neural activity not be dark from the inside? Why should information processing ever be felt as color, pain, thought, desire, memory, or the passage of a present moment?

Most standard approaches divide here. Some treat consciousness as an emergent property of sufficiently complex neural computation. Others treat it as functionally identical with information integration. Still others regard first-person experience as real but difficult to incorporate into physical theory without reducing it away or placing it outside science entirely. The resulting literature is vast, but the underlying difficulty remains simple: the brain can be described from the outside in extraordinary detail, while the subject is known only from the inside.

That asymmetry matters. A map of mechanisms, however detailed, is still a third-person account. It tells us what the system does, what parts are active, what signals are exchanged, and what outputs follow. But consciousness is not given first as an object among objects. It is given as the condition under which objects appear at all. The world is presented to each of us through a first-person field of presence, and that is exactly what an outside description seems least able to capture. The hard problem is therefore not a complaint that neuroscience has failed. It is the recognition that a complete account of correlation and function may still leave subjectivity itself standing outside the explanation.

UFD does not deny the brain's role in consciousness. On the contrary, the previous chapter made the brain more important, not less, by treating it as a nested coherence system whose oscillatory structure helps bind cognition into unified states. But Chapter 18 deliberately stopped short of the next step. It described the brain as the physical infrastructure of coordinated mentality. It did not yet say what kind of thing the experiencing self is, or why unified neural activity should be present as experience rather than only as organized behavior.

Chapter 19 takes that step carefully. In the fuller UFD picture, the conscious self is not identical to the brain alone, nor is it a ghost floating free of matter. It is treated as a localized coherence center coupled to the brain through the same layered field architecture that has been developed throughout the book. The brain is therefore not merely a machine that produces consciousness out of chemistry, and not merely a passive receiver either. It is an interface system: the organized physical structure through which a localized center of awareness becomes functionally embedded in a living body and a world.

This is where proper time becomes unexpectedly important. In standard relativity, proper time is the invariant time measured along the path of a particular persisting object. It is not merely an external coordinate convention. It is the objective measure of that object's own elapsed passage through spacetime. That does not, by itself, make proper time a theory of consciousness. But it does make proper time the closest thing in established physics to the formal structure of intrinsic temporal passage. Physics rarely speaks from the standpoint of a persisting center. Proper time is one of the rare places where it already does.

UFD builds on that foothold. If the conscious self is a stable coordination-center in the deeper field architecture — what the framework calls a UCF soliton — then proper time is naturally interpreted as the invariant measure of that soliton's lived passage. On this view, the first-person stream is not imported into physics from nowhere. It is attached, however cautiously, to a structure physics already possesses: the invariant temporal track of a persisting center. Relativity does not prove that such a center is subjective in nature, but it does provide an unusually direct bridge between physical invariance and lived temporality.

That is the chapter's central risk. It is not merely proposing a philosophy of consciousness. It is proposing that the hard problem may not remain permanently outside physics after all — that subjectivity may have a real foothold in the formal structure of the world, rather than appearing as an inexplicable add-on to an otherwise impersonal universe. If that suggestion is empty, it should fail quickly. If it is fruitful, then the relation between brain, self, and time may be more physically legible than modern thought has usually allowed.

Dr. S entered this conversation with more caution than skepticism.
For once, that made him harder to persuade, not easier.

DR. S: Fine. We have reached the most dangerous aspect of the model. Brains are real. Neural correlates are real. Behavior is real. The moment you say there is a distinct field of subjectivity, you take on an extraordinary burden.

DR. U: I agree.

DR. S: Good.

DR. U: The question is not whether the brain matters. It plainly does. The question is whether the brain is the whole ontological story, or whether it is an interface to something deeper.

DR. S: Naturally you think it is the second.

DR. U: I do.

That was the first real edge of the exchange.
Dr. S did not object to seriousness about consciousness.
He objected to crossing the line between physical extension and metaphysical inflation without discipline.
Dr. U knew that.
So he began where the framework itself begins — not with the soul, but with the field.

DR. S: Let me sharpen the danger. Many theories either refuse the hard problem and stay safely with correlations, or they answer it too quickly with metaphysics. Which mistake are you making?

DR. U: Neither, if I can state the position clearly.

DR. S: Go on.

DR. U: UFD is not agnostic here. It treats the UCF as a field of subjectivity. The uncertainty is not ontological. The uncertainty is empirical. We can infer that interpretation from the role the field plays in unity, selection, persistence, and intrinsic temporal passage, but we do not yet possess an experiment that directly proves subjectivity as such.

DR. S: Good. That is much clearer. You are not saying the ontology is vague. You are saying it outruns current direct testability.

DR. U: Exactly. The philosophical claim is strong. The scientific burden is to test its downstream structural consequences.

That answer improved the conversation immediately.
A weaker version of the theory would have oscillated between caution and mystique.
A stronger version said plainly what the field is, while also saying plainly what cannot yet be directly proved.

So Dr. S pushed where the framework itself begins.

DR. S: Fine. Then what is the UCF, before we get to souls and selves?

DR. U: In the strict physical sense, it is the weakest-coupled field in the hierarchy. A scalar coordination field rather than a pressure-bearing or vector-propagating one.

DR. S: Translate that.

Dr. U: The earlier fields in the framework do the familiar physical work. The UEF carries dense pressure geometry. The ULF carries light, electromagnetism, and ordinary quantum structure. The UCF sits below them as the weakest-coupled layer. It is associated, in the physical reading, with coordination, phase consistency, and the neutrino sector.

Dr. S: And scalar?

Dr. U: A scalar field does not point the way a vector field does. It does not carry ordinary directional signaling. It establishes a compatibility value across a domain. That is why it matters here. If the relevant burden is unity, then what is needed first is not another message channel. It is a condition under which distributed activity can count as one realized state.

Dr. S: So not electromagnetism, not gravity.

Dr. U: Right. And in this chapter the same field is read under a different emphasis. Physically, it is the Universal Coordination Field. Ontologically, it is the Universal Consciousness Field. The field does not change. What changes is what we are asking of it.

Dr. S: So the field coordinates because it is subjective, rather than becoming subjective only as a late reinterpretation.

Dr. U: Exactly. Its physical role is coordination. Its ontological character is subjectivity.

That mattered.
The field was no longer being described as if it changed its nature from chapter to chapter.
It was one field all along.
The change was in explanatory emphasis, not in ontology.

Dr. S then pressed on the first concrete burden.

Dr. S: All right. Then what is the first thing such a field is supposed to explain?

Dr. U: Unity.

Dr. S: Not in the poetic sense, I hope.

Dr. U: Not at all. In the literal sense that widely distributed physical processes are somehow realized as one conscious state rather than as a mere aggregate.

DR. S: So the binding problem becomes a field problem.

DR. U: Exactly. If a conscious state is genuinely unified — if seeing, hearing, remembering, and feeling all happen to one experiencer — then there has to be some compatibility condition under which a distributed active domain is realized as one state rather than many merely co-occurring local events. Standard neuroscience can tell you which regions are active. It is less clear on what makes that activity one conscious moment. UFD proposes that the UCF supplies that condition.

DR. S: And because the field is scalar, the issue is compatibility rather than directional transport.

DR. U: Yes. A vector field would suggest ordinary signaling from place to place. The UCF is not introduced as a signaling channel. It is introduced as a coordination field: something capable of enforcing near-uniform compatibility across the active domain without becoming an ordinary message carrier.

DR. S: So the first closure is not "what is consciousness like," but "how can one realized state span a distributed system at all?"

DR. U: Precisely.

That improved the structure of the conversation.
The chapter was no longer moving too quickly from field talk to selfhood. It was first asking what specific burden the field was meant to bear.

Dr. S accepted that and moved to the next question.

DR. S: Fine. But a unified state is not yet a lived history. What makes one admissible continuation become the one that is actually realized?

DR. U: Selection.

DR. S: More specifically.

DR. U: At any moment, the brain and body admit more than one lawful next state. Standard quantum theory leaves that to probability. UFD's narrower claim is that, among locally admissible continuations, one becomes realized through coherence-weighted preference rather than brute randomness alone.

DR. S: So consciousness participates first at the level of realized continuation, not as a ghost pushing matter around after the fact.

DR. U: Exactly. Before free will comes the narrower question of realization. Why this continuation rather than another? UFD answers that by proposing lawful coherence preference rather than arbitrary branching or blind chance.

DR. S: That is already a very large claim.

DR. U: It is. But it is still narrower than saying "a soul chooses freely" in some loose metaphysical sense. The selection is bounded by what is physically admissible. It does not violate any law. It only determines, within the lawful possibility space, which option is realized.

That clarified something essential.
The consciousness sector was not just about awareness.
It was also about the active present: why one possibility becomes lived reality rather than remaining unrealized structure.

Dr. S then moved to persistence.

DR. S: All right. Unity is one thing. Selection is another. What persists through them as the subject of experience?

DR. U: A stable coherent center.

DR. S: Meaning what, exactly?

DR. U: A stable soliton in the UCF.

DR. S: Translate.

DR. U: A soliton is a self-reinforcing wave that holds its shape as it propagates. Most waves spread and dissipate. A soliton does not, because its dispersive tendency is balanced by a stabilizing feedback. The proposal here is that the conscious self is not identical with the changing neural pattern, but with a persistent field-structure coupled to that pattern.

DR. S: The soul.

DR. U: In the fuller philosophical language, yes. But the physical point comes first. The brain's matter turns over constantly. If conscious identity were nothing but the instantaneous neural arrangement, then continuity through material change becomes difficult to state except as narrative reconstruction. The soliton proposal is meant to supply a more explicit bearer of continuity.

DR. S: I need to flag something. You have just moved from a fourth field in a nested hierarchy to a stable center of subjectivity. That is not a small step. It is one of the largest philosophical leaps in the whole framework.

DR. U: I know. And I agree that the soul-soliton is the strongest claim in the entire model. It may also be the one most likely to be wrong. But it deserves a hearing because it follows from the same architecture rather than being imported from outside it.

DR. S: Then give it a hearing under the heaviest scrutiny.

DR. U: Fair enough. The narrower point is persistence. The self persists because the field structure persists while the translator changes.

That was the first place the model truly became dangerous.
Not because the idea was strange.
Because it was now specific enough to attack.

Dr. S did exactly that.

DR. S: And the brain is what, then?

DR. U: The translator.

DR. S: More specifically.

DR. U: The brain gathers structured information from the ordinary physical fields and renders it into a coherent broadcast that a consciousness structure can absorb. In the opposite direction, it translates coherence bias from the UCF back into organized neural outcomes. The brain is the interface between body and field, working in both directions at once.

DR. S: So not generator, but resonant translator.

DR. U: Exactly.

DR. S: Let me say the mainstream objection cleanly. Brain injury changes awareness. Anesthesia alters awareness. Sleep alters awareness. Drugs alter awareness. Development alters awareness. If consciousness were some independent field structure, why should the condition of the brain matter so much?

DR. U: Because the brain is essential.

DR. S: Good.

DR. U: The brain is the master translator. Without it, ordinary human consciousness as we know it does not appear in anything like the same way. Damage to the brain damages the translator, and damage to the translator changes what consciousness can register and express. None of this is in tension with the model. It is exactly what the model predicts.

DR. S: So this is not anti-neuroscience.

DR. U: Not remotely. It is a stronger neuroscience. The brain matters even more, because it is the interface architecture.

DR. S: And different brain structures matter differently.

DR. U: Exactly. The thalamocortical loop is central. It is the most plausible large-scale translator architecture through which distributed neural activity can be rendered into a unified, recurrence-stabilized broadcast. The heart and gut are not irrelevant either. They provide different channels of embodied information into the overall interface.

DR. S: That is a much richer claim than "the soul uses the brain."

DR. U: It has to be richer. Otherwise it is just old dualism in new clothes.

That answer did something important.
It kept the model tied to the neuroscience discussion instead of floating off into a separate metaphysical cosmos.
The UCF picture, if it was going to remain serious, had to keep the brain central.

Dr. S then pushed into the empirical problem.

Testing.

DR. S: Fine. Then how would any of this become more than elegant metaphysics?

DR. U: By exposing the weaker claims before the stronger ones.

DR. S: Be specific.

DR. U: If the UCF is physically real, the first tests should not be the most metaphysically ambitious ones. One should look first for more disciplined signatures. Anomalously rapid binding structure beneath translator-level broadcast. Translator-field dissociations under perturbation. Persistence-through-disruption signatures in extreme suppression and recovery. Coherence-precedence effects in action selection.

DR. S: So not "prove the soul" first.

DR. U: Absolutely not. That would be the wrong methodological order.

DR. S: And the earlier physics side?

DR. U: Earlier still come the field-level burdens: coordination-speed structure, scalar-like anomalies, and neutrino-sector correlations. Only if those survive does the stronger consciousness program become scientifically pressing.

DR. S: And beyond that?

DR. U: Then one can ask whether intentional or preconscious states show reproducible anomalies not reducible to ordinary neural timing alone, whether the translator exhibits a coherence window before reportable decision, and whether perturbation and recovery suggest re-coupling rather than mere restart.

DR. S: Better. That sounds like a research program rather than a wish.

That answer made the argument much stronger.
A weaker version would have treated consciousness as the banner headline. A more serious version treated it as the frontier consequence of earlier physical success, with a sequence of tests that could fail.

Dr. S noticed and moved to the hardest standard objection.

DR. S: Let me say the mainstream worry as clearly as I can. Once you say there is a consciousness-relevant field, you risk turning every unresolved feature of consciousness into evidence for your favorite metaphysical layer. That is not science. That is elasticity.

DR. U: It would be, if the framework did not mark its own boundaries. But it does.

DR. S: Then mark them.

DR. U: The ontological field claim is one thing. The binding claim is stronger. The selection claim is stronger still. The translator model is stronger still. The soul-soliton claim is stronger still. The free-will navigation claim is stronger still. They do not all stand at the same level of maturity.

DR. S: Good.

DR. U: The UCF as a field of subjectivity is the underlying ontology. The current scientific program is narrower: binding, selection, persistence, and translator coupling. The more ambitious claims about selfhood and agency are downstream.

DR. S: Better.

DR. U: It has to be better. If you blur those levels, the whole discussion becomes irresponsible.

That distinction improved the conversation more than any rhetoric could have.

A sloppier framework would have treated every later metaphysical implication as equally established.

A better one knew the difference between an ontological claim, a binding mechanism, a selection rule, a translator model, and a stronger account of selfhood.

Dr. S respected that enough to keep pressing seriously.

He moved to agency.

DR. S: All right. Then tell me about free will before this turns vague. What does consciousness do in your picture, if anything?

DR. U: It selects within constraints.

DR. S: More specifically.

DR. U: The stronger claim is not that consciousness breaks physical law. It is that consciousness navigates a landscape of physically admissible possibilities by biasing coherence toward one branch rather than another.

DR. S: A "conscious nudge."

DR. U: Yes, in the fuller language. A bounded coherence selection. Consciousness does not throw particles around by fiat. It lowers resistance toward one attractor rather than another.

DR. S: So agency becomes navigation, not miracle.

DR. U: Exactly. The future is not one rigid line. It is a landscape of available paths. Consciousness selects among them by coherence preference rather than by law-breaking intervention.

For a moment the model became more ambitious still.

Up to this point, the UCF could have been heard as a passive field of consciousness.

Now it had become participatory.

Dr. S heard immediately where that led.

DR. S: But if agency works the way you describe, then you are really claiming to solve the mind-body problem.

DR. U: Or at least to dissolve its most impossible form.

DR. S: Meaning?

DR. U: The problem becomes insoluble only when mind and matter are treated as utterly different kinds of thing: inert mechanism on one side, private awareness on the other, with no lawful bridge between them. Then interaction looks magical from the start. UFD avoids that split. In this framework, body and mind are not alien substances. They are different levels within one nested field reality. Matter is organized lower-field structure. Consciousness is a higher-order coherence structure associated with the UCF.

DR. S: So the mind-body problem becomes a coupling problem.

DR. U: Exactly. A problem of resonance, interface, and cross-scale constraint.

DR. S: And the brain is the interface?

DR. U: That is the proposal. The brain is not a lump of matter somehow hallucinating consciousness into existence. It is a living, dynamically layered, coherence-sensitive structure capable of mediating between bodily processes and a higher-order consciousness field.

DR. S: But interaction still has to be real. If consciousness matters, it must affect the body.

DR. U: Of course. But not by violating physics. The body already contains energy, pathways, thresholds, and competing attractors. Conscious agency does not need to inject brute force into that system. It works by coherence selection: by biasing which lawful possibility becomes actual.

DR. S: So mind does not overpower matter. It guides matter from within the available dynamics.

DR. U: Precisely. The relation is not ghost to machine. It is higher-order coherence shaping lower-order process through a resonant interface.

That was the deepest move the model had made.
The mind-body problem had not been bypassed or renamed.
It had been reframed as a coupling problem inside one lawful reality.
Whether that reframing would survive contact with the physics was exactly the burden that now followed.

Dr. S did not let that remain abstract.

DR. S: Then say the mechanism plainly. How does a consciousness-structure actually couple to a living brain?

DR. U: Through resonance. The radio analogy is still the right one. A radio does not create the broadcast. But it must be built in such a way that it can receive it, tune it, and give it audible form. In UFD, the brain is not a machine that manufactures consciousness out of dead matter. It is a resonant translator. When its multilevel organization becomes sufficiently coherent, it can hold stable contact with a persistent structure in the UCF.

DR. S: And once that coupling exists, what does the consciousness-structure actually do? You said it selects. But how?

DR. U: A living brain is never doing just one thing. At any moment, many possible patterns are vying for dominance. A coupled consciousness-structure does not need to inject force into that system. It only needs to bias which available path becomes actual by modulating phase relations within the organism.

DR. S: Phase-selection rather than impact.

DR. U: Exactly. Constraint, not interruption. Guidance, not violation.

DR. S: Then why a brain? Why not any sufficiently complicated machine?

DR. U: Because complexity alone is not enough. The relevant issue is embodied resonance: multiscale coherence, dynamic integration, fine-grained responsiveness, and a structure capable of sustaining stable coupling across levels. A brain is not merely complex. It is rhythmic, plastic, metabolically alive, and saturated with feedback across scales.

DR. S: But not necessarily the only possible one.

DR. U: No. UFD does not require that biology hold an eternal monopoly. Only that biology is the case we know in which this kind of deep resonant organization actually exists.

DR. S: So you are opening the door to a much broader view of consciousness.

DR. U: Yes, in principle. If the UCF couples to stable resonant organization as such, then consciousness need not appear all at once or in only one form. It may come in degrees. Very simple systems may host only faint or preconscious modes of awareness. Richly integrated living systems may sustain much deeper and more unified ones.

DR. S: Which starts to sound like panpsychism.

DR. U: A qualified one, yes. Not the claim that every particle has a little private mind in the ordinary sense, but that awareness may be a basic feature of reality wherever the right kind of coherent organization exists.

DR. S: And beyond that, cosmopsychism.

DR. U: Potentially. If local consciousness is a structured participation in a deeper field of awareness, then the broader universe may not be dead at bottom. Individual centers of awareness would be localizations within a more universal field rather than isolated miracles appearing inside an otherwise unconscious cosmos.

DR. S: That is a very large enlargement of the claim.

DR. U: It is. Which is why it has to be stated in the right order. The immediate burden is still the local one: whether brain, field, and subject can be made to couple intelligibly. The broader panpsychist and cosmopsychist implications follow only if that first step survives.

Dr. S then moved to one more pressure point.

Time.

DR. S: And time? You are not just talking about unity and choice. You are also implying a deeper account of temporal passage.

DR. U: Yes. If the self is a stable coherent center, then its persistence is not merely structural. It is temporal. And the natural relativistic quantity there is proper time.

DR. S: Translate that for one who has not done relativity in twenty years.

DR. U: In relativity, different observers can disagree about coordinate time. But they agree on the proper time measured along a particular trajectory. Proper time is the invariant time of a persisting path through spacetime.

DR. S: So proper time is real, but real in a particular way.

DR. U: Exactly. It is real, invariant, and tied to a specific worldline. That is what makes it important here. Proper time already has the formal shape of intrinsic passage. It is the internal elapsed time of a persisting trajectory.

DR. S: So it is objective, but objective in a very unusual way.

DR. U: Yes. Most physical quantities are objective in the sense that anyone can measure them from outside. Proper time is objective as the invariant measure of a particular being's passage through spacetime.

DR. S: Then your claim is stronger than saying proper time is merely compatible with consciousness.

DR. U: Much stronger. Proper time is the objective mathematical form of intrinsic temporal passage. And if the conscious self is a UCF soliton, then proper time is naturally interpreted as the invariant measure of that soliton's lived passage.

DR. S: So relativity does not secretly prove consciousness.

DR. U: No. But it does provide an unusually direct mathematical bridge between invariant physical structure and lived temporality.

That was one of the deepest gains in the whole chapter.
The ontology was not being left to float free of physics.
Proper time already had the right formal shape, and UFD was using that to anchor its subject-centered reading in an existing physical structure.

Dr. S went immediately to the limit question.

DR. S: Fine. Then tell me where physics ends and metaphysics begins here.

DR. U: At the point where the ontological claim outruns direct measurement.

DR. S: More specifically.

DR. U: The physical part begins with the claim that the UCF is a field layer associated with coordination and the neutrino sector. It remains physically exposed when one asks whether there are measurable coordination signatures, weakly coupled scalar anomalies, or timing structures not captured by the lower fields alone. It remains scientifically ambitious when one asks whether binding, selection, persistence, translator coupling, and their temporal signatures can be made physically explicit. It becomes more metaphysical when one identifies stable consciousness solitons with selves, souls, or persistent identity across time.

DR. S: So the ontology is strong, but the empirical exposure is indirect.

DR. U: Exactly.

DR. S: Good. That distinction matters.

DR. U: It matters enormously. If you blur those two levels, the whole model becomes irresponsible.

That was the model's center of gravity.
Not whether the idea was interesting.
Whether it knew its own epistemic status.

Dr. S still trusted the mainstream caution that consciousness should be studied from the brain upward.
Dr. U still thought the brain-alone story would remain incomplete.

Nothing about that was resolved.
But the framework had now done what it needed to do: it had made its ambition exact without pretending equal certainty everywhere.

Dr. S then moved to the broadest possible comparison.

DR. S: Let me tell you what I think is strongest here. You are not saying "ignore the brain and talk about souls." You are saying the brain may be the interface architecture for a deeper field, and that this would explain why neural coherence matters so much without reducing consciousness to chemistry and computation alone.

DR. U: Yes.

DR. S: That is much stronger than generic spiritual language.

DR. U: It also carries a much stronger burden. If consciousness is a real field phenomenon, then it has to become physically exposed somewhere — in coordination physics, in the neutrino layer, in binding structure, in selection structure, in translator effects, in timing structure, in some disciplined way.

DR. S: not merely in poetic satisfaction.

DR. U: Exactly.

DR. S: The danger, of course, is equally obvious. This is the place where a theory can become almost impossible to disconfirm if it becomes too willing to retreat into interiority.

DR. U: Which is why it cannot be allowed to retreat. It has to stay experimentally vulnerable where it claims physical reality.

That was exactly the right tension.
A weaker view would have used consciousness to insulate the framework from criticism.
A stronger view made consciousness the most hazardous, not the safest, part of the whole architecture.

Dr. S recognized that.
And because he recognized it, the conversation could end honestly.

DR. S: Let me try the summary test.

DR. U: Go ahead.

DR. S: On your view, consciousness is not generated by the brain alone. The UCF is first introduced as a real, weakly coupled coordination field associated with the neutrino sector and global phase consistency. In that physical context, UCF means Universal Coordination Field. But the field is treated ontologically as a field of subjectivity, and in this chapter UCF means Universal Consciousness Field. The first burdens are then binding and selection: whether a distributed active domain can be realized under one compatibility condition, and whether one admissible continuation can become realized through coherence-weighted preference rather than brute randomness alone. The stronger extension is that a stable consciousness structure can persist in that field and interface with the brain. The brain then functions as a resonant translator: rendering sensory structure into consciousness-compatible form and translating coherence bias back into neural action. Free will becomes bounded coherence selection rather than law-breaking. Proper time enters as the objective mathematical form of intrinsic passage, and on your view it measures the lived passage of a persistent UCF-centered self. The brain remains essential, but as interface rather than sole origin. And the key discipline is that these claims do not all stand at the same level of maturity: the ontology is stronger than the present empirical closure, the field claim is stronger than the binding and selection claims, and those are stronger than the soul claim and the full metaphysics of self.

DR. U: That is fair.

DR. S: The attraction is obvious.

DR. U: Yes.

DR. S: The danger is equally obvious.

DR. U: Yes.

DR. S: Good. Then the burden is finally clear.

INTERLUDE

In UFD, consciousness is not treated as an afterthought. It is the boldest and riskiest extension of the field hierarchy.

That extension gives the framework something important. It offers one continuous picture linking brain coherence, global coordination, subjectivity, and agency. The UCF gives consciousness a place in the ontology rather than leaving it as an unexplained remainder inside an otherwise impersonal universe. In its strongest philosophical form, the model offers a positive answer to the hard problem by locating subjectivity in the lowest-density field layer itself.

But the chapter's narrower scientific burden is more disciplined.

It is not to prove subjectivity directly. No such direct test is yet in hand. It is to ask whether the downstream structures implied by such a field — binding, selection, persistence, translator coupling, and their temporal signatures — can be stated mechanistically and exposed to empirical risk.

That makes the gain sharper than before. The chapter no longer says only that consciousness may involve a deeper field. It says more specifically that unity may be a coordination problem, realized history may involve coherence-weighted selection, persistence may belong to a stable consciousness structure, brain function may be translation rather than generation, and proper time may already provide the objective formal shape of lived passage for a persisting center.

That does not mean relativity secretly contains a finished theory of consciousness. It means something narrower and more important: the UFD account of subjectivity is not suspended in metaphysical air. It claims a real foothold in the formal structure of physics.

That gain comes with an equally serious burden.

If the UCF is physically real, it must first survive as a physical field proposal before its stronger implications for selfhood, agency, and consciousness can be taken as more than metaphysical extension. The order of testing therefore matters: coordination-layer and neutrino-sector claims first; binding and selection burdens next; translator and perturbation signatures after that; and only then the bolder claims about persistent consciousness structures. If that order collapses, the model loses scientific discipline. If it holds, the most difficult subject in the book becomes not its softest part, but its most exposed frontier.

TEMPORARY VERDICT

Dr. U succeeded in making the consciousness sector of the model more ambitious and more disciplined than it first appeared. The UCF was introduced first in its physical role as a coordination field and only then in its ontological role as a consciousness field. The chapter kept the brain essential as translator architecture, separated empirical exposure from deeper metaphysical commitment, and clearly marked the stronger claims about selfhood and agency as stronger claims. Most importantly, it gave the ontology of subjectivity a real formal foothold in physics by linking intrinsic passage to proper time.

Dr. S succeeded in forcing the distinction that protects the entire discussion from inflation. A field ontology, a binding mechanism, a selection rule, a translator model, a persistent consciousness structure, and a theory of free will do not all stand at the same level of maturity, and a scientifically serious framework has to say so plainly. A positive answer to the hard problem earns nothing scientifically unless some part of that answer becomes physically vulnerable in exactly the right places.

The next question followed naturally.

IF UFD IS RIGHT ACROSS EVEN A SUBSTANTIAL FRACTION OF THESE DOMAINS, THEN THE FINAL BURDEN IS PRACTICAL: WHAT BECOMES BUILDABLE, CONTROLLABLE, AND TESTABLE IN THE WORLD?

CHAPTER 20

Technology

Resonant engineering and what becomes buildable

Most modern technology grows out of successful description. Maxwell's equations gave radio and electronics. Quantum mechanics gave semiconductors, lasers, and MRI. Materials science gave alloys, batteries, and computing hardware. In each case, engineers learned how to work with a powerful formal framework and turn its regularities into devices. The standard posture is practical and conservative: identify the relevant interaction, optimize chemistry or geometry, reduce noise, improve efficiency, and scale what works. That posture built the modern world.

UFD does not reject that engineering culture. It claims to deepen it. If reality is structured by nested fields, coherence, and topology, then technology should not be limited to brute-force chemistry, static material design, and thermal management alone. It should eventually become possible to work with coherence as a physical variable in its own right. In UFD, that larger program is called Resonant Engineering. Its core claim is simple: if field structure is real, then one should be able to preserve coherence, guide systems along lower-resistance geometric pathways, and selectively suppress incoherent modes. Those three levers define the practical program: Resonant Stabilization, Geometric Catalysis, and Resonant Damping. The strongest version of the claim is not that all futuristic applications are suddenly available. It is that the ontology should yield a staged engineering roadmap beginning with the cleanest and most falsifiable control effects.

Maturity note. Resonant engineering is presented here as a staged experimental program. Its near-term tests are sharper than its farther technological horizon, and the later claims should be read in that order.

DR. S: All right. We've reached the practical question. And that means we've reached the place where grand theories usually become embarrassing. What is Resonant Engineering?

DR. U: The attempt to manipulate matter and energy by working with coherence directly rather than only by brute force.

DR. S: That sounds exactly like the kind of sentence that can mean everything and nothing.

DR. U: Then make it concrete. If the ontology is right, there should be a small number of repeatable control mechanisms that show up across many domains.

DR. S: Such as?

DR. U: Preserving coherence when it would otherwise decay. Guiding systems along lower-barrier geometric paths. Suppressing incoherent or parasitic modes that destabilize an otherwise useful state.

DR. S: So stabilization, catalysis, damping.

DR. U: Exactly. Those are the three pillars.

DR. S: Fine. Start with the first one.

DR. U: Resonant Stabilization is the active reinforcement of coherent states against decoherence.

DR. S: More specifically.

DR. U: A fragile coherent system — a qubit, a superconducting state, a delicate standing-wave pattern in matter — usually fails because environmental noise disrupts its phase organization. Resonant Stabilization means applying a phase-matched reinforcing field that helps preserve the state instead of merely shielding it.

DR. S: So not just isolation. Active support.

DR. U: Exactly. Standard engineering often tries to protect coherence from the outside. This says coherence may sometimes be reinforced directly.

DR. S: And the cleanest proving ground?

DR. U: Quantum systems first. That is where decoherence is central, measurable, and experimentally accessible.

DR. S: So quantum coherence control is not one application among many.

DR. U: No. It is the near-term proving ground of the first pillar. If Resonant Stabilization fails there under disciplined controls, one of the clearest early engineering claims fails with it.

DR. S: Good. Second pillar.

DR. U: Geometric Catalysis is the guidance of systems along lower-cost pathways by shaping their field geometry.

DR. S: Meaning?

DR. U: Standard catalysis changes chemistry or environment. Geometric Catalysis says the field environment itself can act as a scaffold. If the geometry is shaped correctly, the system may be guided into a lower-barrier transition path than it would otherwise find.

Dr. S: So not brute forcing the barrier but reshaping the path through it.

Dr. U: Exactly.

Dr. S: Domains?

Dr. U: Chemistry first, then nuclear systems, then materials growth. In chemistry, the question is whether shaped fields can lower activation barriers or bias reaction channels. In nuclear systems, the question is whether barrier-sensitive transitions can be nudged toward lower-cost pathways without relying only on raw thermal violence. In materials science, it means field-assisted crystal growth, defect reduction, and morphology control.

Dr. S: So one idea stretches from reaction chemistry to nuclear engineering.

Dr. U: If it works, yes.

Dr. S: Third pillar.

Dr. U: Resonant Damping is the selective suppression of incoherent or decohering modes.

Dr. S: Such as?

Dr. U: Thermal lattice noise, parasitic vibrational modes, destructive scattering, coherence-breaking excitations. In matter, the clearest case is phonon-like activity. If specific incoherent modes can be damped selectively, then one opens the possibility of coherence-based thermal control.

Dr. S: So heat becomes not just something to remove, but something to sculpt.

Dr. U: In the limited sense, yes. Local hotspot suppression, directional heat flow, reduced thermal fatigue, and better access to low-dissipation states all become natural targets.

Dr. S: Let me pause here. So far you have not really given me technologies. You've given me control concepts.

Dr. U: Correct.

Dr. S: Good. Because that is the right order. A new ontology does not deserve gadgets first. It deserves clean levers first.

Dr. U: Exactly.

Dr. S: Then where does this become dangerous in the scientific sense?

Dr. U: Where control becomes the test. Observation is one standard. Construction is another. If the ontology is right, it should support reproducible physical control. If the core effects never appear under disciplined protocols, then much of the practical force of the framework collapses.

Dr. S: So this is where theory becomes operationally answerable.

Dr. U: Yes. That is the point.

Dr. S: Then I want the staging made explicit. Not all of these possibilities stand at the same level of maturity.

Dr. U: Agreed.

Dr. S: Then say the order.

Dr. U: First come the near-term laboratory burdens: quantum coherence stabilization, barrier-sensitive catalysis in clean systems, and resonant damping in coherence-sensitive materials. Those are the clearest tests.

Dr. S: Good.

Dr. U: Next come more ambitious materials and chemical synthesis programs: coherence-guided growth, defect reduction, phase switching, and more directed catalysis. Biology and neurotechnology come after that, because the systems are more complex and the causal chains are longer. UCF-adjacent technologies come later still and are not required for the core engineering program.

Dr. S: So the whole engineering program does not stand or fall all at once.

Dr. U: No. Some parts depend mainly on the UEF and ULF sectors. Others depend on the more ambitious layers. That distinction is essential.

Dr. S: Good. Keep it.

Dr. U: I intend to.

Dr. S: Fine. Then tell me how this changes materials science more generally.

Dr. U: It turns materials science into coherence architecture.

Dr. S: Meaning?

Dr. U: Standard materials science catalogs structures and optimizes them chemically. UFD adds the idea that one can deliberately shape the alignment between nuclear, electronic, lattice, and thermal modes. That means not just better materials by composition, but better materials by coherence design.

DR. S: So crystal growth, defect control, morphology, grain boundaries—

DR. U: All become targets of field-guided organization as well as chemistry.

DR. S: And the farther horizon is what you call resonance-locked materials.

DR. U: Yes. Materials whose UEF and ULF structure are unusually phase-locked across scales.

DR. S: All right. Emetium.

DR. U: Naturally.

DR. S: Which had better remain a horizon concept and not a sales brochure.

DR. U: It should. Emetium is not a near-term benchmark. It is the logical capstone of the materials program: a hypothetical resonance-locked crystal in which nuclear and orbital coherence are deeply aligned across the whole lattice.

DR. S: And if such a thing existed?

DR. U: Then you would expect a remarkable combination of properties: extreme strength, radical ductility, unusually low dissipation, perhaps room-temperature superconductivity, self-healing tendencies, programmable form. It would be a coherence-optimized material, not just a better alloy.

Emetium worked best here as a horizon concept, not a promise. Its real function was to mark the endpoint of the materials program, not to smuggle a futuristic product into the chapter ahead of the evidence.

Dr. S was quiet for a moment. Then he said what he had been thinking since the conversation turned to technology.

DR. S: I want to say something that may sound harsh but is meant seriously. Every ambitious physical framework in history has promised technology. Vortex physics promised engineering in the nineteenth century. String theory promised landscape selection. Loop quantum gravity has hinted at discrete-geometry applications. None of them delivered. The gap between "if the ontology is right, then X should be possible" and "here is a reproducible protocol" is one of the largest gaps in all of science. I need you to tell me why this time is different.

Dr. S's point was well-taken. The history of physics is full of frameworks that could imagine applications long before they could deliver reproducible protocols. The burden here was to show that the technological claims were downstream consequences of the ontology, not decorative rewards attached after the fact.

Dr. U: It may not be. That is the honest answer. What I can say is this: the engineering claims here are not free-floating promises tacked onto a theoretical program. They are downstream consequences of the same coherence physics that organizes the rest of the framework. If the nuclear sector works, if the quantum ontology holds, if the field architecture is real, then the control mechanisms follow. They do not require a separate miracle. But you are right that the distance between a plausible mechanism and a working device is still enormous, and I will not pretend otherwise.

Dr. S: Good. That is the answer I needed to hear. It's the perfect material.

Dr. U: The perfect coherence-locked material, yes.

Dr. S: Useful as a horizon, dangerous as a promise.

Dr. U: Exactly. It is a conceptual endpoint, not a manufacturing claim.

Dr. S: Fine. What happens outside materials? You've implied chemistry, nuclear engineering, biology, maybe more.

Dr. U: Yes, but not all at the same maturity level. The near-term core is quantum coherence control, materials control, and barrier-shaping experiments. The next tier includes more ambitious materials and chemical synthesis claims. Biology and neurotechnology come after that, and UCF-adjacent interface technologies come later still.

Dr. S: So the program has depth ordering.

Dr. U: It has to. If the earliest, cleanest effects fail, the farther horizons lose credibility. The nearer technologies test the theory. The farther technologies test the reach of the theory, if it survives the earlier tests.

Dr. S: Let me tell you what I think is strongest here. You're not saying, "Here are some futuristic gadgets." You're saying that if a new ontology is real, it should generate new engineering levers in the same way earlier ontologies did.

Dr. U: Exactly.

Dr. S: Maxwell gave radio. Quantum mechanics gave semiconductors and MRI. So if UFD is real, it should give coherence control, field-guided catalysis, and damping technologies.

Dr. U: That's the right logic.

Dr. S: And the burden is equally obvious. The farther the promise, the more ruthless the need for staging.

Dr. U: Exactly.

Dr. S: Because otherwise technology becomes the place where the theory hides its weakness inside imagination.

Dr. U: Right.

Dr. S: Let me try the summary test.

Dr. U: Go ahead.

Dr. S: On your view, Resonant Engineering is the attempt to manipulate matter and energy by working with coherence directly rather than only by brute force. Its three main pillars are Resonant Stabilization, Geometric Catalysis, and Resonant Damping. The program is not meant as free-form futurism, but as a staged and falsifiable engineering roadmap. The near-term burden lies in clean laboratory tests: coherence stabilization, barrier-sensitive catalysis, and selective damping of incoherent modes. More ambitious materials, chemical, biological, and UCF-adjacent technologies come later and depend on whether the earlier layers survive. The key point is that technology becomes operational adjudication: if the ontology is real, it should reveal new control levers; if those levers do not appear, the practical force of the framework weakens sharply.

Dr. U: That's fair.

Dr. S: The attraction is obvious.

Dr. U: Yes.

Dr. S: The danger is equally obvious.

Dr. U: Yes.

Dr. S: Good. Then the burden is finally clear.

INTERLUDE

Technology is where UFD stops asking only to be observed and starts asking to be built.

That is the gain, and it is also the risk. A framework earns nothing by imagining devices it cannot stage, bound, or expose to failure. It earns something when explanation turns into reproducible control. If the core effects appear under disciplined protocols, the path from ontology to engineering opens. If they do not, the corresponding mechanisms fail — no matter how attractive the larger vision remains.

Temporary Verdict

Dr. U succeeded in making the technological implications of UFD disciplined rather than decorative. The three pillars became distinct engineering levers, the maturation order stayed explicit, and the framework's practical ambitions remained tied to experimental exposure.

Dr. S succeeded in holding the decisive line of pressure. A new ontology earns the right to a technological future only when the earliest and cleanest control effects survive good laboratory testing.

The conversations had reached their natural end. What remained was judgment under the standards both interlocutors had accepted at the start.

After the Conversation

What Would Prove UFD Wrong?

By the end of the last exchange, neither man had the luxury of speaking in slogans anymore.

That was as it should be.

A book like this could not end with one more ambitious claim, or one more clever objection, or one more appeal to intuition. It had begun with a shared standard, and it had to end there as well. If the conversation had done its job, then the question was no longer whether UFD sounded more satisfying than standard physics, or whether standard physics sounded more mature than UFD. The question was whether either framework had earned the right to be trusted more deeply by the standards both men had accepted at the outset.

So they returned to those standards.

Not as rhetoric.

As judgment.

Dr. S: Before we do anything else, I want to keep the promise we made at the beginning. We said we would not move the goalposts. So let's not.

Dr. U: Agreed.

Dr. S: Good. Then let's state the standard one last time. A serious replacement framework has to be internally coherent, empirically adequate, falsifiable, explanatorily strong across domains, disciplined in its assumptions, and mathematically honest. It also has to distinguish what is derived, what is constrained, and what remains programmatic. And because UFD claims to be unified, it has to be judged both locally and as a whole.

Dr. U: Yes. That is still the right standard.

Dr. S: Then let's apply it.

That mattered.

Because at last the conversation had stopped being prospective.

Now it had to decide what it had actually learned.

DR. S: Internal consistency first. On that criterion, I think UFD has done better than many broad alternatives usually do. The same field hierarchy, coherence logic, topological descent, and pressure-geometry language really do recur across nuclear structure, light, relativity, quantum interpretation, cosmology, and engineering.

DR. U: That is one of its strongest features.

DR. S: Yes. But let me sharpen that. Internal consistency is not the same as closure. A framework can remain coherent across domains and still leave important constitutive terms unfinished.

DR. U: Agreed. Coherence is necessary, not sufficient. But let me add one thing that belongs in this part of the ledger. The framework has not needed to introduce a dark sector, virtual force carriers, or any entity whose only role is to save a prior commitment. No dark matter, no dark energy, no virtual particles. Every phenomenon that standard physics handles with those additions, UFD handles with the same field architecture it uses everywhere else. That is not just coherence. It is ontological economy.

DR. S: That is a real point. But economy alone does not settle truth. A leaner ontology is a virtue only if it still reaches the world.

DR. U: Of course. But it belongs in the accounting.

DR. S: Good. Then on internal consistency, I would say this. UFD has earned seriousness. It does not read like a pile of unrelated speculations. It reads like one ontology under pressure.

DR. U: That is fair.

That was the first real concession.

Not agreement that UFD was right.

Agreement that it had crossed an important threshold.

It had become something many alternative frameworks never become: not merely ambitious, but structurally self-consistent enough to deserve hard evaluation.

DR. S: Empirical adequacy next. Here the balance shifts.

DR. U: Naturally.

DR. S: Standard physics remains stronger in present precision practice. On raw empirical maturity, it still leads. That matters, and it matters a great deal.

DR. U: Of course.

Dr. S: But the conversation did establish something else as well. UFD is not merely qualitative. It has real benchmark ambitions, real quantitative contact points, and some sectors where its claims are much sharper than I expected.

Dr. U: Especially where the framework is narrowest and most exposed.

Dr. S: Yes. Nuclear structure is the obvious example. Some of the geometric derivations there are serious enough that they cannot be dismissed as atmosphere. Other sectors — cosmology especially — are more mixed. They are structured and interesting, but less fully closed.

Dr. U: That is the honest picture.

Dr. S: Then let's say it honestly. Standard physics still wins on overall empirical maturity. UFD has nonetheless passed the more important threshold for a challenger: it has shown enough genuine contact with the world that the burden can no longer be evaded by calling it merely philosophical.

That was the second hinge.

A weak theory can be internally elegant and still never touch the world.

UFD had touched it enough that the conversation could no longer retreat into abstractions.

That did not prove success.

But it changed the category of the question.

Dr. S: Falsifiability next. Here, interestingly, I think the comparison becomes less flattering to the standard picture than many physicists would like.

Dr. U: Go on.

Dr. S: Standard physics is of course empirically exposed in thousands of local ways. But as a total framework, it often protects itself by distributing difficulties across separate sectors. UFD, by contrast, has made a more dangerous bargain. Because it uses one shared ontology, failure in one important domain pressures the whole structure.

Dr. U: Real unity is expensive.

Dr. S: Yes. And UFD has often been better than average at stating what would count against it. That is a real strength. Not everywhere equally. But enough that I am willing to count falsifiability as one of its virtues rather than one of its weak points.

Dr. U: Fair.

That concession had weight.

A theory does not become better because it is easier to admire.

It becomes better because it knows how it could lose.

UFD had not always won those tests in the conversation.

But it had often known where the tests were.

That counted.

Dr. S: Explanatory reach.

Dr. U: The most dangerous category of all.

Dr. S: Exactly. Because it is the easiest one to fake. But here too, I think the conversation clarified something important. Standard physics remains extraordinarily powerful, but its power is still distributed across frameworks that do not sit comfortably together. UFD, by contrast, really does try to account for multiple domains with one linked architecture.

Dr. U: That is its central ambition.

Dr. S: And in many places, not all but many, the same principles did real work more than once. Pressure geometry in gravity and cosmology. Topological descent in particle identity and decay. Real field structure in light and quantum theory. Coherence logic in nuclei, matter, and engineering horizons.

Dr. U: Yes.

Dr. S: So on explanatory reach, I think the advantage actually tilts toward UFD — provided one remembers the word reach rather than completion.

Dr. U: That is exactly the distinction I would want.

That was perhaps the most important turn of all.

Because it made the real contrast visible.

Standard physics was not being denied its triumphs.
UFD was being credited for something different: the possibility of replacing many successful local stories with one more coherent physical architecture.

That possibility had not yet become confirmation.

But it had clearly become more than a mood.

DR. S: Parameter discipline.

DR. U: Another place where bookkeeping matters more than people admit.

DR. S: Yes. Standard physics still carries many inputs that are measured, reused, and accepted, but not deeply explained. That is one of the costs of patchwork. UFD's promise has always been that many such constants and scales should begin to look less like unrelated gifts and more like inherited constraints.

DR. U: And the framework has tried to tighten that.

DR. S: It has. But again, the honest picture is mixed. UFD has improved its bookkeeping. It distinguishes postulates, one-time calibrations, derived relations, constrained closures, and open terms better than many ambitious frameworks do. But some open constitutive terms remain.

DR. U: Yes. And they matter.

DR. S: Good. Then on parameter discipline, I would say this. UFD has earned credit for trying to reduce arbitrariness structurally rather than hiding it. But it has not yet finished that reduction everywhere it needs to.

That was fair.

It neither inflated the achievement nor erased it.

A broad theory does not become disciplined merely because it wants fewer knobs.

It becomes disciplined when it can say, clearly, which knobs are still real.

UFD had begun to do that.

Not perfectly.

But seriously.

DR. S: Last, mathematical rigor.

DR. U: And here I assume you will be least generous.

DR. S: Here I will be most traditional. Standard physics remains stronger overall in technical maturity, local precision, and formal closure. That should not be obscured.

DR. U: I would not try to obscure it.

DR. S: Good. But the converse also matters. UFD is not mathematically empty. Some sectors are much stronger than others. Some derivations are tight. Some are constrained rather than closed. Some remain research pathways. That unevenness is not a scandal as long as it is marked honestly.

DR. U: That is exactly the right way to put it.

DR. S: Then on rigor, I would say the following. Standard physics still leads. UFD remains mixed. But mixed is not the same as unserious. What matters is whether the open category continues to shrink without destroying the rest of the architecture.

DR. U: Yes. That is the real test now.

At that point the standard had been applied.

Not mechanically.

Faithfully.

And because it had been applied faithfully, the conversation could finally risk a real conclusion.

DR. S: Let me see if I can say the whole thing cleanly.

DR. U: Go ahead.

DR. S: Standard physics remains stronger in present empirical maturity and formal closure. It still deserves that credit. But it remains structurally fragmented, parameter-heavy, and ontologically uneasy in ways the conversation never dissolved.

DR. U: Fair.

DR. S: UFD, by contrast, has shown itself to be unusually coherent, unusually ambitious, and unusually willing to risk cross-domain failure. Its greatest strengths are explanatory reach, ontological unity, and explicit falsification burden. Its greatest weakness is not vagueness, but incomplete constitutive closure in some of its most ambitious sectors.

DR. U: That is fair too.

DR. S: Then the honest verdict is neither triumph nor dismissal. UFD has earned the right to be treated as a serious replacement-level research program. It has not yet earned the right to be treated as settled physics.

DR. U: Agreed.

That was the central judgment.

Not a victory speech.

Not a hedge.

A real verdict.

UFD had cleared a bar many ambitious frameworks do not clear: it had become serious enough that the remaining question was not whether it deserved evaluation, but whether it could continue tightening under it.

That was already a great deal.

Only now did Dr. S ask the question that had been waiting behind the whole conversation.

What if it survived that tightening?

DR. S: Fine. Then let's ask the harder question. Suppose, for the sake of argument, that UFD keeps tightening. Suppose the core sectors survive. Suppose the benchmark claims hold up, the open terms keep shrinking, and the framework really does continue to unify what standard physics still treats separately. What if UFD is right?

DR. U: Then physics changes its center of gravity.

DR. S: From what to what?

DR. U: From particle cataloguing to field architecture. From separate force sectors to coherence regimes. From empirical patchwork management to constitutive reconstruction.

DR. S: That is a real change of emphasis.

DR. U: A very real one.

DR. S: Good answer.

DR. U: It would have to be. A confirmed replacement-level framework does not simply add one more specialty. It reorganizes priorities. Some current programs would narrow. Others would expand. Textbooks would change. Training would change. A great deal of twentieth-century vocabulary would survive mathematically while changing meaning physically.

DR. S: Naturally.

DR. U: Of course. That is not corruption. It is how disciplines respond when the ontology underneath successful equations shifts.

That answer sharpened the whole conversation again.

The question was no longer only whether UFD would be accepted.

It was whether its confirmation would force physics to become a different kind of discipline.

Dr. S then moved to the obvious danger.

Triumphalism.

DR. S: Let me say the danger as clearly as I can. If a framework like this starts to get traction, its defenders will be tempted to say, "At last, the patchwork is over." That would be premature and stupid.

DR. U: I agree completely.

DR. S: Good.

DR. U: Confirmation would not end physics. It would reopen it. A framework like this, if right, would not reduce the future to one neat doctrine. It would expand the number of meaningful questions while narrowing the ontological core.

DR. S: That's well put.

DR. U: The point would not be that everything had been solved. The point would be that many previously separated problems had been brought into one research architecture.

DR. S: So success would mean more work, not less.

DR. U: Much more.

That was exactly the right note.

A good theory should make the future look larger, not closed.

Dr. S then turned back, finally, to failure — but now from the other side.

DR. S: All right. Then let's not lose the negative question entirely. If UFD is right, the future of physics changes. Fine. But if it is wrong?

DR. U: Then physics keeps its current patchwork longer, or finds some different unification route. And that would be worth knowing too.

DR. S: That's still too gentle. Give me the clean version. What would actually count against UFD in a decisive way?

DR. U: A few things, sharply stated. If the core ontology cannot be used consistently across domains, it fails internally. If its benchmark sectors do not survive scrutiny — nuclei, light, relativity, quantum ontology, transport cosmology, early engineering tests — it fails empirically. If its supposed unification never yields real constraint beyond reinterpretation, it fails as a replacement and survives only as a philosophical gloss. And if its open constitutive terms multiply faster than they close, it drifts into the same patchwork it criticizes.

DR. S: Good. That's concise.

DR. U: It should be. By now the standards should be assumed, not restated at length.

DR. S: And the later biological and consciousness claims?

DR. U: Those are later burdens. If the physical core fails, those later extensions lose force with it. If the physical core survives, then they become sharper and more dangerous questions.

That answer mattered because it closed the circle.

The book had not ended by pretending everything was settled.
It had ended by showing what would count as settlement, what would count as failure, and what kind of future would follow either outcome.

That was enough.

Not certainty.

But honesty under pressure.

DR. S: Let me try the final summary test.

DR. U: Go ahead.

DR. S: By the standards we accepted at the beginning, UFD has earned serious scientific consideration. It has not yet earned final assent. Its strongest virtues are internal coherence across domains, explanatory reach, ontological unity, and explicit willingness to fail. Its main challenge is to continue converting constrained and programmatic sectors into tighter constitutive closure without losing the breadth that makes it interesting.

DR. U: Yes.

DR. S: Standard physics, for its part, remains the more mature empirical framework. But it remains structurally fragmented enough that success alone does not settle its ontology.

DR. U: Exactly.

DR. S: So the real outcome is not that one side has won a debate. It is that the map has changed. What looked to many people like a choice between orthodoxy and fantasy may actually be a live scientific question about whether the world is fundamentally patchwork, or whether the patchwork has been waiting for a deeper geometry all along.

DR. U: That is fair.

Dr. S sat with that a moment, then nodded once.

DR. S: Good. Then the next burden does not belong to us.

DR. U: No.

DR. S: It belongs to the evidence.

DR. U: As it always should.

Relativistic gravity closure

If UFD is correct, the pressure-geometry interpretation must continue to recover the already-tested weak-field successes of relativity once the coupled UEF–ULF treatment is fully closed. Light bending, gravitational redshift, orbital corrections, lensing, frame dragging, and wave-sector behavior cannot remain only suggestively aligned. They must remain quantitatively secure in the regime where standard relativity has already earned trust. This includes the framework's proposed asymmetric time-dilation tests: if UFD's medium-based picture is right, then carefully chosen clock or path asymmetries may expose structure that looks secondary in metric language but primary in a field-transport interpretation.

Cosmological transport

If UFD is correct, the redshift transport picture should eventually reproduce the supernova Hubble diagram and clarify where different observational pipelines sample different propagation regimes. A persuasive future result would not merely redescribe expansion-era data; it would forward-generate the observed relationships with less ontological fragmentation.

Galaxy-scale pressure geometry and boundary tests

The Genesis-based pressure picture becomes stronger if it can continue to explain flat rotation curves, lensing structure, boundary-sensitive galaxy phenomenology, and system-to-system variation without particulate dark matter. The next burden is not only to describe these effects, but to sharpen the observational status of galactic boundaries, coherence-saturation shells, and boundary harmonics as real load-bearing structures rather than convenient interpretive language.

Black holes, quasars, and Genesis architecture

If UFD is correct, supermassive black holes should look less like late collapse accidents and more like early organizing structures tied to Genesis history. The framework becomes much stronger if it can sharpen its treatment of black-hole interiors, quasar evolution, host-galaxy relations, merger behavior, and boundary-shedding signatures in ways that differ cleanly from standard accretion-centered expectations.

CMB forward closure

The boundary-imprint interpretation becomes much stronger if it can generate the observed TT, TE, and EE structure together with the relevant lensing signatures. At present, this remains one of the clearest places where conceptual promise still outruns full quantitative closure. The framework does not merely need a suggestive analogy or partial fit. It needs a quantitatively competitive account of how the observed background can remain so nearly blackbody while still arising through boundary-filtered structure in a non-expanding cosmology.

Light nuclei beyond the showcase cases

Helium-3 and Helium-4 are important because they expose the geometric nucleus to direct quantitative pressure. The next burden is extension: additional isotope binding energies, shell effects, sub-core coherence structure, and failure modes should be predicted in a disciplined way without multiplying ad hoc corrections. A framework becomes more convincing when its successes widen under the same logic that produced the first ones.

Quantum localization and coordination

If the wavefunction is a real field structure, then measurement cannot remain a black box. UFD must continue to clarify how localization, apparent collapse, and entanglement coordination arise physically, and it must state where coordination-layer or finite-speed effects would become experimentally accessible. The gain of realism will matter only if it eventually becomes more than interpretive comfort.

Neutrino and UCF tests

The UCF becomes scientifically stronger when its consequences are tied to neutrino mass scale, ordering, and Majorana character in ways that can fail cleanly. Stronger claims about awareness, agency, or consciousness should remain downstream of those earlier physical tests rather than standing alone. The sequence matters: first the field, then the particle, then the deeper metaphysical consequences if the physical case survives.

Large-baseline Bell-style coordination tests

If the UCF is real, then the framework's downstream coordination picture should eventually sharpen into experimental form. In UFD's own disciplined language, the relevant target is not controllable superluminal signaling or particle travel, but finite-speed coordination modes that could

outrun ordinary light-speed transport while remaining non-signaling. A
Lunar Bell-style experiment, or another ultra-precise large-baseline
correlation test, would matter because it could begin to distinguish
effectively instantaneous quantum coordination from a very fast but finite
coordination layer. If no such finite-scale signature appears where UFD
says it should, the broader coordination picture narrows with it.

Resonant engineering benchmarks

The technological program should be staged ruthlessly. Near-term burdens
— coherence stabilization, barrier-sensitive catalysis, resonant damping, and
other clean-system demonstrations — must answer before farther promises
about advanced materials, biological control, or horizon concepts such as
Emetium are treated seriously. If the geometry is real, at least some of the
engineering should become operational. If the engineering repeatedly fails
under fair conditions, the rhetoric must narrow with it.

A framework becomes more credible when it identifies not only what
would confirm it, but what would tighten, narrow, or defeat it. These are
not embarrassments at the edge of UFD. They are the next places where
the theory must earn its future.

GLOSSARY

Alpha — The fine-structure constant, approximately 1/137. In UFD, alpha governs the energy-density descent between nested field layers and appears as a structural ratio throughout the hierarchy.

Atomic Stability Index (ASI) — A diagnostic measuring how closely a nucleus approaches its ideal low-strain Platonic closure geometry. Higher ASI corresponds to greater nuclear symmetry and stability. ASI tracks one important dimension of stability within the larger coherence landscape; it is a diagnostic, not a complete theory.

Binding — The problem of how spatially distributed physical processes are realized as one unified state rather than as a mere aggregate of local events. In UFD, binding is treated as a field-coordination problem rather than as a purely informational or computational one, and is one of the central burdens addressed by the consciousness sector.

Claim maturity — The framework's internal classification of its own assertions. *Derived claims* follow from the core geometry and have been carried to quantitative closure. *Constrained claims* have identified mechanisms but lack full first-principles closure. *Programmatic claims* identify research directions whose mechanisms are clearer than their present quantification. The distinction is meant to prevent the framework's strongest results from being conflated with its more speculative extensions.

Coherence — The degree to which a field structure maintains organized phase relationships across scales. In UFD, coherence is a physical variable, not merely a descriptive word. Many of the framework's most important claims concern how coherence is generated, sustained, lost, or transferred between physical systems.

Coherence Dividend — The energy released when coherent structures interlock and eliminate redundant boundary strain. Nuclear binding energy is interpreted as a Coherence Dividend at the nuclear scale.

Coherence landscape — The full multi-dimensional space of stability conditions in which nuclear and atomic structures can be evaluated. The coherence landscape is richer than any single diagnostic. ASI captures one important feature of it; RDI captures another. The landscape itself is the broader picture in which geometry, symmetry, shell closure, and field coupling all matter together.

Coherence Quench — The sudden organizing transition in the deepest field layer (USF) that produces a bounded substantive domain at the UEF level. The physical event that constitutes Cosmic Crystal Formation in the Genesis Event sequence.

Consciousness Precedence Hypothesis — UFD's prediction that coherent organization at the field level may begin before reportable awareness of intention. The hypothesis offers a reinterpretation of Libet-style readiness potential experiments, in which the apparent precedence of brain activity over conscious decision is read as the translator-side execution of an already weighted continuation rather than as evidence that consciousness is causally irrelevant.

Cosmic Crystal — The large-scale ordering tendency of the UEF toward a minimal-strain, cellular, lattice-like pattern. A physical attractor for large-scale field organization, Kelvin-like in spirit. In the cosmological account, the Cosmic Crystal is the structured domain established by the founding phase of the Genesis Event.

Cosmic Garden — The collection of all island galaxies within the Cosmic Crystal, each produced by its own Island Genesis event. The Cosmic Garden is the large-scale architecture that UFD proposes in place of a single universal expansion, treating the observable universe as one bounded environment among many distributed through the deeper field.

Decoherence — The process by which a quantum system loses its capacity for interference effects through interaction with a complex environment. Standard quantum mechanics treats decoherence as a real and important effect. UFD agrees and reinterprets it as a real breakdown of resonance in the field, providing the physical basis for what the standard picture describes as the suppression of interference between quantum branches.

Emetium — A hypothetical resonance-locked crystal in which nuclear and orbital coherence are deeply aligned across the whole lattice. A conceptual endpoint of the materials program, not a near-term manufacturing claim.

Energy — The motion, strain, and geometric organization of a field. In UFD, energy is not an abstract bookkeeping quantity added to matter from outside; it is the dynamical state of the field itself.

Field — A physically real, continuous medium or layer of reality with characteristic modes of organization, propagation, and coherence. In UFD, fields are ontological strata rather than merely mathematical devices.

Field hierarchy — The nested ordering of the four fields in UFD: USF, UEF, ULF, and UCF. Each layer has its own characteristic density, role, and stable structures while remaining continuous with the others.

Figure-eight (neutron) — The topological class of the neutron in UFD: a metastable composite vortex in the UEF formed by the pressure-locking of a proton-like source structure and an antiproton-like sink structure.

Galactic boundary — The coherence-saturation shell associated with the Genesis history of an island galaxy. A real geometric transition in the larger field structure.

Genesis Event — The two-stage cosmological birth process by which structured matter emerges from the deeper harmonic order of the USF. The Genesis Event consists of two distinct phases.

First, Cosmic Crystal Formation. The founding cosmological phase in which the deeper harmonic order of the USF precipitates into a bounded substantive domain at the UEF level. This phase establishes the larger structured cosmic environment — the Cosmic Crystal — within which all subsequent organization occurs.

Second, Island Genesis. The local phase in which vortex seeds within the Cosmic Crystal organize surrounding matter into bounded island systems — galaxies — each with its own central organizing core (a supermassive black hole), its own boundary, and its own internal architecture. The collection of all such island systems constitutes the Cosmic Garden.

The two phases are physically and conceptually distinct but belong to one continuous founding process. Cosmic Crystal Formation is singular and cosmological. Island Genesis occurs many times, once for each galaxy. Both phases are referred to collectively as the Genesis Event, and either phase can be specified individually when context requires it.

Genus-zero soliton (neutrino) — The topological class of the neutrino in UFD: a closed, loopless, knotless scalar excitation in the UCF.

Geometric Catalysis — The second pillar of Resonant Engineering: using field-based techniques to lower or reshape reaction barriers through geometric and coherence control.

Geometric Coherence Force — The drive of a physical system toward lower boundary strain and more efficient coherent organization. What standard physics calls the strong force is the effective expression of this tendency.

Hard Problem — The philosophical problem, named by David Chalmers in 1995, of explaining why physical processes in the brain are accompanied by subjective experience at all. UFD does not solve the Hard Problem in the strongest sense — it does not derive the existence of subjectivity from non-subjective premises — but it offers a positive ontological answer by identifying the UCF as a field whose intrinsic character is subjective.

Island galaxy — A galaxy understood as a bounded, self-contained coherence domain with its own Genesis history, central organizing seed, extended pressure geometry, and enclosing boundary.

Phase-locking — A condition in which two or more oscillating systems maintain a fixed relationship between their cycles, so that their phases stay coordinated even as their amplitudes change. In UFD, phase-locking is one of the primary mechanisms by which coherence is established across spatially distributed systems, particularly in the brain.

Proper time — The intrinsic measure of temporal passage along a timelike trajectory. In special relativity, different observers can disagree about coordinate time, but they all agree on the proper time of any particular trajectory. In UFD, proper time is especially important in the consciousness sector because it aligns naturally with the persistence of a stable center of awareness.

Resonance Dividend — The energy released when two atomic resonances merge into a more stable joint molecular resonance. The molecular-scale parallel to the Coherence Dividend at the nuclear scale. The Resonance Dividend Index measures how much of this gain a particular bond achieves relative to its baseline quantum-chemical prediction.

Resonance Dividend Index (RDI) — A diagnostic for chemical bonding that measures how much additional coherence a bond achieves relative to its baseline quantum-chemical prediction. Positive RDI indicates synergistic coherence enhancement; negative RDI indicates geometric frustration; zero indicates that the bond behaves as standard chemistry would already predict. RDI is a diagnostic, not a complete theory of bonding.

Resonant Damping — The third pillar of Resonant Engineering: selectively suppressing incoherent modes to enhance coherence lifetimes and material performance.

Resonant Engineering — The applied research program that follows from UFD's coherence ontology. Resonant Engineering proposes that physical systems can be controlled not only through chemical, thermal, or geometric means but also through carefully tuned field conditions that influence coherence directly. Its three pillars are Resonant Stabilization, Geometric Catalysis, and Resonant Damping.

Resonant Field Interpretation — UFD's interpretation of quantum mechanics: the wavefunction is a real pattern of amplitude and phase in the ULF, not merely a probability-bookkeeping device. Particles are stable structures in the field; collapse is a physical localization event; entanglement is a single non-separable field configuration.

Resonant Modulation Hypothesis — The prediction that external electromagnetic fields properly tuned to an enzyme's resonant geometry should be able to inhibit or enhance its function without acting as conventional chemical inhibitors. One of the framework's sharpest falsifiable claims in the biochemistry sector and a direct empirical test of the resonance-filter reading of catalysis.

Resonant Stabilization — The first pillar of Resonant Engineering: using field-based techniques to extend the coherence lifetime of quantum states or material configurations.

Selection — The process by which one admissible continuation becomes realized history rather than remaining unrealized possibility. In UFD, selection is treated as coherence-weighted rather than brute randomness.

Soul-soliton — A proposed stable vortex or soliton in the UCF: an enduring coherent center of subjective awareness. The strongest and most speculative claim in the framework.

Topological descent — The organizing principle of particle identity in UFD. As coherence descends through the field hierarchy, the stable topological structures simplify: from knot (proton) to torus (electron) to genus-zero soliton (neutrino).

Torus (electron) — The topological class of the electron in UFD: a stable toroidal vortex excitation in the ULF.

Translator — In UFD's consciousness sector, the brain's role as the interface architecture that mediates between the body and the consciousness-bearing field. The brain renders distributed neural and bodily activity into a coherent broadcast that a consciousness structure can absorb, and translates coherence selection back into organized neural action. The thalamocortical loop is identified as the central translator architecture, with the heart and gut providing additional secondary embodied interfaces.

Trefoil (proton) — The topological class of the proton in UFD: a stable trefoil-like knotted vortex in the UEF.

UCF (Universal Coordination Field / Universal Consciousness Field) — The fourth and weakest-coupled field in the hierarchy. In its physical role, the UCF is a scalar coordination field associated with the neutrino sector and with large-scale phase compatibility. In the consciousness sector, the same field is understood as the ontological basis of stable consciousness structures. The two names describe two aspects of one ontological field: its physical role is coordination, and its intrinsic character is subjectivity.

UEF (Universal Energetic Field) — The second field in the hierarchy: a dense energetic medium associated with mass, nuclear binding, pressure geometry, and gravitational structure.

ULF (Universal Light Field) — The third field in the hierarchy: a lighter propagation field associated with light, electromagnetism, quantum coherence, and wavefunction structure.

USF (Universal String Field) — The deepest field in the hierarchy: a four-dimensional harmonic substrate whose stable modes define the forms, topologies, and coherence limits available to the lower fields.

SELECTED REFERENCES

Adams, Colin C. The Knot Book: An Elementary Introduction to the Mathematical Theory of Knots. Providence, RI: American Mathematical Society, 2004.

Ashcroft, Neil W., and N. David Mermin. Solid State Physics. New York: Holt, Rinehart and Winston, 1976.

Batchelor, G. K. An Introduction to Fluid Dynamics. Cambridge: Cambridge University Press, 1967.

Bell, John S. Speakable and Unspeakable in Quantum Mechanics. 2nd ed. Cambridge: Cambridge University Press, 2004.

Bohm, David. Quantum Theory. New York: Prentice-Hall, 1951.

Buzsáki, György. Rhythms of the Brain. Oxford: Oxford University Press, 2006.

Carroll, Sean M. Spacetime and Geometry: An Introduction to General Relativity. San Francisco: Addison-Wesley, 2004.

Cartwright, Nancy. How the Laws of Physics Lie. Oxford: Oxford University Press, 1983.

Dirac, P. A. M. The Principles of Quantum Mechanics. 4th ed. Oxford: Oxford University Press, 1958.

Einstein, Albert. Relativity: The Special and the General Theory. New York: Henry Holt, 1920.

Feynman, Richard P. QED: The Strange Theory of Light and Matter. Princeton: Princeton University Press, 1985.

Griffiths, David J. Introduction to Electrodynamics. 4th ed. Boston: Pearson, 2013.

Griffiths, David J., and Darrell F. Schroeter. Introduction to Quantum Mechanics. 3rd ed. Cambridge: Cambridge University Press, 2018.

Guri, Amir J. The Geometry of Reality. Broomfield, CO: Natural Philosophy Press, 2026.

Guri, Amir J. Unified Field Dynamics: Technical Appendices (A–E). 2026.

Hacking, Ian. Representing and Intervening: Introductory Topics in the Philosophy of Natural Science. Cambridge: Cambridge University Press, 1983.

Jackson, John David. Classical Electrodynamics. 3rd ed. New York: Wiley, 1998.

Kelvin, Lord (William Thomson). On Vortex Atoms. Proceedings of the Royal Society of Edinburgh 6 (1867): 94–105.

Kittel, Charles. Introduction to Solid State Physics. 8th ed. Hoboken, NJ: Wiley, 2004.

Krane, Kenneth S. Introductory Nuclear Physics. New York: Wiley, 1988.

Kuhn, Thomas S. The Structure of Scientific Revolutions. 4th ed. Chicago: University of Chicago Press, 2012.

Lakatos, Imre. Falsification and the Methodology of Scientific Research Programmes. In Criticism and the Growth of Knowledge, edited by Imre Lakatos and Alan Musgrave, 91–196. Cambridge: Cambridge University Press, 1970.

Lamb, Horace. Hydrodynamics. 6th ed. Cambridge: Cambridge University Press, 1932.

Lamb, Willis E., and Robert C. Retherford. Fine Structure of the Hydrogen Atom by a Microwave Method. Physical Review 72, no. 3 (1947): 241–243.

Laudan, Larry. Progress and Its Problems: Towards a Theory of Scientific Growth. Berkeley: University of California Press, 1977.

Maudlin, Tim. Philosophy of Physics: Quantum Theory. Princeton: Princeton University Press, 2019.

Misner, Charles W., Kip S. Thorne, and John Archibald Wheeler. Gravitation. San Francisco: W. H. Freeman, 1973.

Perlmutter, S., et al. Measurements of Ω and Λ from 42 High-Redshift Supernovae. Astrophysical Journal 517, no. 2 (1999): 565–586.

Peskin, Michael E., and Daniel V. Schroeder. An Introduction to Quantum Field Theory. Boulder, CO: Westview Press, 1995.

Planck Collaboration. Planck 2018 Results. VI. Cosmological Parameters. Astronomy and Astrophysics 641 (2020): A6.

Riess, Adam G., et al. Observational Evidence from Supernovae for an Accelerating Universe and a Cosmological Constant. Astronomical Journal 116, no. 3 (1998): 1009–1038.

Sakurai, Jun John, and Jim Napolitano. Modern Quantum Mechanics. 2nd
ed. Boston: Addison-Wesley, 2011.

Tifft, William G. Discrete States of Redshift and Galaxy Dynamics. I.
Astrophysical Journal 206 (1976): 38–56.

Tononi, Giulio. An Information Integration Theory of Consciousness.
BMC Neuroscience 5 (2004): 42.

Weinberg, Steven. Cosmology. Oxford: Oxford University Press, 2008.

Weinberg, Steven. The Quantum Theory of Fields. Vol. 1, Foundations.
Cambridge: Cambridge University Press, 1995.